PRAISE FOR **THIS IS T**

"Exemplary . . . quickly charges off in surprising and consistently fascinating directions. I did not, for example, expect: national elocutionists, Kim Kardashian's vocal fry, the last Vatican castrato, the telling silence of lizards, Alexa or the delightful, data-based revelation that humans can reliably hear a smile. . . . In the equally illuminating and entertaining second half of the book, Colapinto wades into the hotter waters of gender, sexuality, race and politics."

—Mary Roach, *The New York Times*

"Not-so-young guy damages his vocal cords belting out tunes in a rock band—and a fascinating book is born. . . . An extraordinary look at the unsung influence of the sounds we make and hear."

—*People* (book of the week)

"This book has the mark of an enduring success, a work that readers will be turning to fordecades, not years. In this fascinating exploration of the new science of speech, John Colapinto teaches us not only about the sounds we make, but the people we are."

—David Frum, *The Atlantic*

"The story of human speech is as compelling as any mystery narrative. Ancient history, evolution, social development, gender, culture, and Neanderthals all play a part in the fascinating story that charts our journey from grunts to eloquent speech."

—Peter Jackson, director of the *Lord of the Rings* trilogy

"[An] excellent, polyphonic study of the human voice, from baby talk to swan song."

—*Nature*

"Fascinating . . . Colapinto's narrative is chock full of information, and is something any curious-minded reader will be glad to have spent time with."

—*Publishers Weekly* (starred review)

"Focused, lucidly written . . . A rich trove of science and contemporary culture."

—*Kirkus Reviews*

"Colapinto learns, enlightens and entertains."

—*BookPage*

"Lots of data, evidence, thoughtfulness, and heart here."

—*Booklist*

"This is an amazing work. I didn't realize until I read it that it's the book I've been wanting to read for thirty years. Everything you might want to know, or have wondered about, or didn't even consider about how we speak and listen, exchange information, and the musicality behind all of it is here . . . and then some. I couldn't put it down."

—Daniel Levitin, author of *This Is Your Brain on Music*

"John Colapinto creates a compelling narrative surrounding this vast and complex topic and investigates what makes the voice uniquely human."

—*Science*

"This astonishing and resonant tour of the vocal cords ends up transporting the reader all over the human body, the brain, and the world beyond. John Colapinto seems to be murmuring in your ear the whole way, cracking jokes, telling stories, sharing fabulous secrets—and singing, too."

—Abigail Tucker, *New York Times* bestselling author of *Lion in the Living Room*

"With prose as clear as a bell, Colapinto takes you into the complexities by which sound becomes voice and voice becomes a central feature of human culture—and a way to discover who you are. Reading it was a real pleasure."

—Ron Rosenbaum, author of *The Shakespeare Wars*

"As Colapinto proves, no human attribute has been more crucial to the evolution of our species than our ability to convey meaning through sound. Colapinto's written voice is resonant with humor, insight, intelligence, and sincerity. Highly recommended."

—Christopher Ryan, coauthor of *Sex at Dawn*

"A prime example of quality popular science, striking that ideal balance between informative and entertaining. It embraces the wide-ranging aspects of its subject matter, digging in wherever necessary and capturing the reader's curiosity. But it's also an engaging read, thoughtful and funny and finely crafted. Books that accomplish this combination are few and far between, but John Colapinto has definitely written one that does just that."

—*The Maine Edge*

ALSO BY JOHN COLAPINTO

As Nature Made Him

About the Author

Undone

Becoming a Neurosurgeon

THIS IS THE

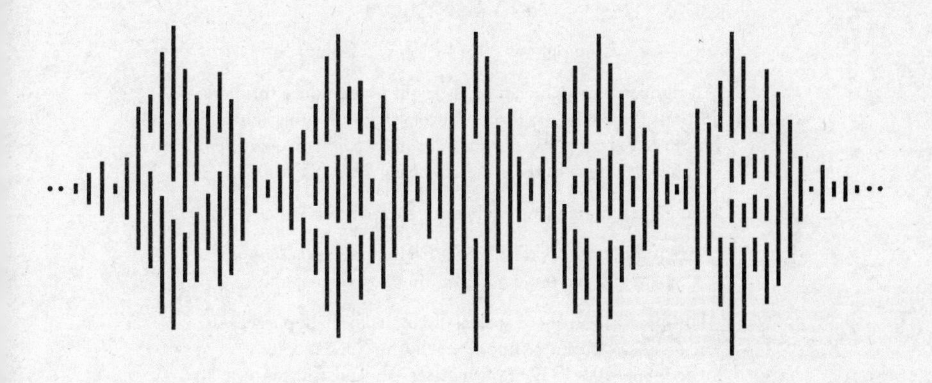

JOHN COLAPINTO

SIMON & SCHUSTER PAPERBACKS

NEW YORK   LONDON   TORONTO   SYDNEY   NEW DELHI

To Donna and Johnny

Simon & Schuster Paperbacks
An Imprint of Simon & Schuster, Inc.
1230 Avenue of the Americas
New York, NY 10020

First Simon & Schuster trade paperback edition January 2022

SIMON & SCHUSTER PAPERBACKS and colophon are registered trademarks of Simon & Schuster, Inc.

Interior design by Ruth Lee-Mui

Manufactured in the United States of America

1 3 5 7 9 10 8 6 4 2

Library of Congress Cataloging-in-Publication Data
Names: Colapinto, John, 1958- author.
Title: This is the voice / by John Colapinto.
Description: New York : Simon & Schuster, [2021] |
Includes bibliographical references and index.
Identifiers: LCCN 2020026734 (print) | LCCN 2020026735 (ebook) |
ISBN 9781982128746 (hardcover) | ISBN 9781982128753 (paperback) |
ISBN 9781982128760 (ebook)
Subjects: LCSH: Voice—Social aspects. | Oral communication—Social aspects. |
Speech and social status.
Classification: LCC PN4162 .C53 2021 (print) | LCC PN4162 (ebook) |
DDC 302.2/242—dc23
LC record available at https://lccn.loc.gov/2020026734
LC ebook record available at https://lccn.loc.gov/2020026735

ISBN 978-1-9821-2874-6
ISBN 978-1-9821-2875-3 (pbk)
ISBN 978-1-9821-2876-0 (ebook)

CONTENTS

INTRODUCTION

PERSONALLY SPEAKING

Some years ago, I was invited by my then-boss, Jann Wenner, the owner of *Rolling Stone*, to be the lead singer in a band he was putting together from the magazine's staff. I had just turned forty-one and I jumped at the opportunity to sustain the delusion that I was not getting old. "Sign me up!" I said. My chief attributes as a singer included impressive volume and an ability to stay more or less in tune, but I was strictly a self-taught amateur. I had, for instance, never done a proper voice warm-up and had certainly never been informed that the delicate layers of vibratory tissue, muscle, and mucus membrane that make up the vocal cords are as prone to injury as a middle-aged knee joint. So, on practice days, I simply rose from my desk (I was finishing a book on deadline and spent eight hours a day writing, in complete silence), rode the subway to our rehearsal space in downtown Manhattan, took my place behind the microphone, jolted my vocal cords from zero to sixty and started wailing over my bandmates' cranked-up guitars and drums.

The folly of this approach became clear to me a few weeks into rehearsals when J. Geils Band front man Peter Wolf, whom Jann had enlisted to perform a song, pulled me aside. "You don't have to sing full out in *rehearsal*, man," he said. "Save something for the show." I followed his advice, but by then my voice had taken on a pronounced rasp. I wasn't concerned. I had suffered hoarseness in the past and it had cleared up. Plus, a little vocal raggedness is never out of place in rock 'n' roll. Also, and perhaps most importantly, I felt no discomfort—so how could I have hurt my throat? I *now* know that an invidious feature of voice injuries is that, when they happen, you feel nothing. The vocal cords have no pain receptors.

I continued attending twice-weekly rehearsals and soon reverted to my old ways—actually singing *harder*, trying to put some of the old volume back into my voice, which was sounding weirdly dampened. I was also finding it difficult suddenly to hit high notes, like the F above middle C in the Stones' song "Miss You" ("*Ohhhhhh,* why'd you have to wait so long?"). Reaching for it, my voice would break up into a toneless rattle, or vanish altogether. This began to concern me as the days ticked down to our gig—a holiday party at a downtown dance club to which Jann had invited two thousand of his closest friends, including a constellation of celebrities (Yoko Ono, Paul Shaffer, Val Kilmer), and hired Cher's soundman to work the mixing board. Singing is as psychological as it is physical. Stress attacks the vocal apparatus, tightening muscles that should remain loose and pliable, restricting breathing, closing off the throat, paralyzing the tongue and lips. I was experiencing all of these symptoms as I took my place, center stage, in the glare of the lights, and began our opening number, the Beatles' song "I'll Cry Instead," originally sung by John Lennon. It would seem a little on the nose to suggest that Yoko and her and John's son, Sean, were looking up at me from the front row, except they were.

Today, I can barely bring myself to listen to the CD of that concert which Jann later presented to each band member as a memento. I wince at the tentative way I sing that *Ohhhhh* in "Miss You," sneaking up on the note from below, sliding into it gingerly. I get there, sort of. But at what cost? By the end of the night, I was growling the lyrics to "White Room" like it was a Tom Waits number.

A three-day bout of laryngitis followed. Then I began speaking in a parched whisper. This eventually "improved" to a torn-sounding rumble. Three months after the gig, I was still speaking as if my words were being stirred through gravel. But I was determined to believe the problem would clear up—until an alarming encounter in the building into which I had just moved with my wife and infant son. Holding open the elevator door for one of my new neighbors, a smiling blond woman, I pointed at the buttons and asked, "What floor?" Her smile vanished.

"You've got a *serious* voice injury," she said.

I demurred, but she cut me off, saying that she was a voice coach who worked with Broadway singers and actors. Only much later did I learn that Andrea wasn't just any voice coach. She was the Executive Director for the Linklater Center of Voice and Language. Kristin Linklater (who died in 2020 at age 84), was the founder of a worldwide network of experts who teach a method of vocal production first described in her 1976 book, *Freeing the Natural Voice*. Along with Stanislavsky's Method, Linklater's system is renowned for having helped liberate actors from the stilted vocal mannerisms of old-school acting, to help them both relax and strengthen the vocal muscles, and thus find their "authentic" voice. And Andrea was having none of my disavowals of serious injury. She said that I had likely damaged one of my "vocal folds" (the technical term for vocal cords), and that she could *see*, in my neck, the compensatory muscle movements I was making as I spoke. I was, she told me, "straining the laryngeal muscles, squeezing them around the vocal folds to help them create sound." She added: "I bet your neck gets pretty sore."

In fact, for weeks I'd been enduring a peculiar sensation in my neck, as if I had scalded the skin.

"You're probably holding tension in other muscles, too," she went on. "Our voice comes from our whole body—from the diaphragm at the bottom of the ribs, to muscles along the spine, to the pelvic floor and psoas muscles, the lower back, the hip sockets. With an injury like yours, you're working harder with all of them. You must be pretty tired by the end of the day." I had been attributing the strange, bone-deep exhaustion that afflicted me every evening to the stresses of new parenthood and finishing my book. Not the muscular effort of *speaking*.

She invited me to drop by her apartment, anytime. She could show me some simple relaxation exercises that would help with the immediate symptoms. I hate presuming on neighbors and knew that I would never avail myself of this kind offer. Andrea shrugged and said: "At the very least, you ought to see a laryngologist, just in case it's . . . something else."

This caught my attention. I grew up in a medical family and was familiar with the euphemism "something else." She meant a tumor. A possible *malignancy*. This had never occurred to me. My rasp was so clearly the result of singing with Jann's band—or was it?

The next day, I arrived at Mount Sinai Hospital. I had an appointment with Dr. Peak Woo, chief of laryngology, a subspecialty of otolaryngology (or ear, nose, and throat medicine) that focuses on the vocal cords. Dr. Woo was a soft-spoken man in his late forties with a kindly bedside manner—the kind of doctor who can grasp the tip of your tongue and pull it slightly from your mouth without it seeming unnatural. With his other hand, he guided down my throat a laryngoscope, a tool that looked like the curved spray attachment on a garden hose, a small light affixed to the end. On a nearby computer screen, the live image of my throat was broadcast, a wet red tunnel at the bottom of which sat my vocal cords: two symmetrical, fleshy, pearly-pink membranes stretched like a pair of lips across the opening of my trachea (or windpipe). Through my open vocal cords I could see the rings of tracheal cartilage descending toward the dark abyss of my lungs. Dr. Woo told me to say "Ahhh." I did so, the membranes swinging together like a pair of drapes across the opening of my windpipe. They furiously vibrated as I produced the sound. They popped open the instant I fell silent. He removed the scope.

It was not, he said, a malignancy.

He pointed to the screen, which held a photo of my vocal cords in the open position. The edge of the left cord was ruler straight. On the margin of the right cord was a small bump. A tumor would be lumpy, asymmetrical. My vocal mass was smooth and regular, as if a tiny pea had been inserted under the semitransparent mucus membrane: a textbook polyp, wholly consistent with my history of over-singing in Jann's band. I had broken a blood vessel in that vocal cord and the unchecked bleeding had created the bump of scar tissue that was interfering with the vocal cord's normal, fluid, rippling action. Sweet pure singing voices are partly the result of vocal cords with clean straight edges that meet flush across the opening of the windpipe as they vibrate. Mine did not, and this is what produced the rasps and rattles and rumbles in my voice.

I asked if he might just snip the offending polyp off in a quick outpatient procedure. Hardly. To have the thing removed I would need to check into the hospital for several days to undergo surgery, which would require not only a general anesthetic but a special paralyzing agent to render me completely immobile—a crucial consideration given the extreme fragility of the vocal cords and the permanent injury to the voice that can result from removing even a micrometer too much healthy tissue. Peering through a high-powered stereo-vision microscope, Dr. Woo would, he explained, reach down my throat with a miniature scalpel mounted on a long, knitting-needle-like extension, slit open the mucus membrane, and use a tiny spoon-shaped tool to "shell out" the mass. Given the outer membrane's gossamer fragility it could not be stitched closed and would have to heal on its own. This would require six weeks of strict postoperative vocal silence.

I left his office with a prescription for a medication to take in the days before the operation. Scheduling the procedure was up to me. He told me to call when I was ready.

I never called.

Why? The usual excuses—no time, too expensive, too risky, and who could afford to stop talking for *six weeks*?—positions easier to maintain with a vocal injury than with almost any other medical emergency. Especially if you don't make your living as a singer, actor, audiobook narrator, news anchor, podcaster, voice-over artist. Like most people who are not voice professionals, I took for granted the sounds that emerged from between my lips, thinking: *As long as I'm getting the words out and being understood, my voice is fine.* Which is not to say that I wasn't self-conscious about my rasp. I would often worry, when meeting someone new, that my growling speech might, erroneously, suggest that I was a two-pack-a-day smoker or a Bukowski-esque barfly. (Fine when I was working for *Rolling Stone* where such behavior was virtually *expected*; less so when, in 2005, I moved to *The New Yorker*.)

Speaking on the phone, which always heightened my awareness of my damaged voice (probably because my speech was being isolated, broadcast back to me through the phone's earpiece), I often worried that

I was conjuring in the brain of my invisible interlocutor the image of a thuggish underworld heavy—a particular concern if I was trying to get a potentially delicate journalistic source to trust me. And it was certainly annoying to pick up the phone and say what sounded to me like a perfectly normal "Hello," and have the person on the other end mistake my crackling, static-riddled voice for my answering machine. There was also the inconvenience of disabusing friends who mistook my rattle as a symptom of the flu. But for all these annoyances and discomforts, I was not (I told myself) *disabled*. I could converse. I could work. By these lights, the surgery was not necessary.

I did, however, take certain measures to preserve what remained of my voice. I tried, for instance, to apply the knowledge Andrea had imparted to me in the elevator; I concentrated on relaxing my neck, stopped *pushing* my voice out with an extra effort of my abdominals. This tended to reduce my volume—or "projection"—but it also eliminated the scalding neck pain and overall exhaustion. I also learned, by unconscious trial and error, to lower my pitch, which seemed to smooth my tone a little. Over time, I was even able to convince myself that the problem had cleared up—a state of denial I sustained *for over a decade*, until a day in late 2012, when I embarked on a new story for *The New Yorker*.[1]

It was about Dr. Steven Zeitels, a vocal surgeon at Massachusetts General Hospital in Boston. Since the mid-nineties, Zeitels had ministered to an array of popular singers—Steven Tyler, Cher, James Taylor—as well as famous TV and radio broadcasters, opera stars, Broadway belters and actors. A few months earlier, he had successfully operated on the British singer-songwriter Adele, removing a vocal polyp that had threatened to end her career. She had thanked him from the stage when collecting several Grammy awards. I had spotted Zeitels's name on *The New Yorker*'s in-house "Master Ideas List"—and while I'm certain that my own vocal malady must have played a subconscious role in my pouncing on an idea that my fellow writers had allowed to languish, my own polyp was far from my mind when I called Zeitels to ask if he might be willing to cooperate with a story. I hadn't even finished my pitch before he interrupted me, saying: "It sounds like you're dealing with a pretty significant vocal issue *yourself*."

Brought up short, I stammered something about having experienced "a little vocal strain" some time ago, and changed the subject. But I could not staunch his clinical curiosity. When I visited Zeitels in Boston for our first set of interviews, he insisted on "looking at" my throat. I hesitated, leery of violating the unwritten ethics of journalism (receiving treatment from a physician could be perceived as a quid pro quo for a favorable story). On the other hand, he was proposing not *treatment*, merely a quick peek, which might be justified on reportorial grounds: it would afford me as intimate a look as possible into Zeitels's methods and manner as a physician—which was, after all, why I was there, shadowing him through his workdays. To say nothing of the fact that, at that stage, I hadn't ruled out writing about my own vocal injury in the piece; Zeitels scoping my throat might make a nice scene. Finally, there was Zeitels's urgent fascination with all aspects of voice injuries. He *wanted* to see my vocal cords.

In short, I had the exam.

Like Dr. Woo, Zeitels peered into my throat with a laryngoscope; he, too, left an image of my vocal cords up on his computer screen. Even to my untrained eye, the mass looked far bigger than in the photo taken more than a decade earlier by Dr. Woo. Zeitels was certainly impressed. "It's like Adele's," he said. "But yours is *magnitudes* bigger. You couldn't possibly sing with something this big. It's mechanically impossible."

He was right about that. The few times I'd tried, my voice shut down, went off-pitch—and the extra exertion of driving air past my burdened vocal cord would force me to reload my lungs at an abnormally fast rate, making my phrasing choppy (good singers time their intakes of breath around natural pauses in a song's lyrics), causing me to hyperventilate and grow light-headed. Little wonder that I had not sung publicly since Jann's party and no longer sang even in private, around the house. Too exhausting. Too depressing.

I missed it, and this gave me some emotional insight during the interviews I conducted with Zeitels's patients, many of them professional performers whose singing voices had been silenced. The most renowned, and notorious, was Julie Andrews, who, in 1997, while performing in a Broadway production of *Victor/Victoria*, suffered hoarseness, was diagnosed with a polyp, and underwent surgery at New York's Mount Sinai (this was some years before Dr. Woo's tenure there). She emerged from the operation not only bereft of the preternaturally pellucid tone that had made her famous, but unable to sing at all without experiencing the rattling, pitch shifts, drop-outs, and dizziness that I knew all about. She successfully sued the hospital—but never got her singing voice back. In 2000, she turned, in desperation, to Zeitels, who tried, in four separate operations, to repair the damage, but in vain. "She'd lost too much vocal cord tissue in the earlier operation," he said, "and much of what remained was stiffened with scar tissue."

For Andrews, who had been performing professionally since age ten, and for whom singing formed an essential part of her identity and livelihood, the loss was devastating. "To feel that that would never come my way again!" she told me with feeling. "The huge joy—apart from singing itself—is the wonder of singing with a very big symphony orchestra. It's ecstasy." Another of Zeitels's patients, a former New York City Opera tenor scheduled to undergo surgery to remove some vocal cord scarring that had ended his singing career, told me why he was hoping to return to professional performing at the unlikely age of forty-nine. He had been working as a singing teacher, but "I've grown a little tired of just *talking* about it," he said. "I mean, when you sing, you're giving voice to your *soul.*" I related to these testimonials, and admitted as much to Zeitels—although I hastened to add that, of course, for me, singing had always been a mere hobby, a pleasant pastime, and that I had no right to compare my meager loss to that of real vocal artists.

"Why not?" he said. "Singing *meant* something to you. Gave you pleasure. Expressed something inside. It's mysterious. People who do it, at any level, report that it has a profound effect on them psychologically, emotionally, spiritually."

He let me know, however, that my singing was not the primary issue. There was also the question of my speaking voice. Yes, I could still *talk*, he said, but my altered voice was affecting my life in ways that I was not acknowledging. "Here's the way to understand your speaking voice," he said. "You're grossly hoarse. People might say, 'Well, his voice isn't that bad.' No. Your voice is actually *pretty bad.* Your right vocal cord—the one with the polyp—has a severely impaired elastic dynamic capability. You're working at three or four percent of normal."

Consequently, he said, I had done what many people with my injury do: I had developed strategies for, as he put it, "speaking around the problem"—retraining my recurrent laryngeal nerve (the nerve that, among other things, controls the tension on the vocal cords) to drop the pitch of my voice, slackening my freighted vocal membrane so that the 3 or 4 percent that was still pliable would vibrate. This reduced the rattle in my voice, but at a cost. It was robbing me of the natural variation in pitch and volume that people use to give color, animation, expression, and personality to their utterances—what linguists call *prosody*, the melody of everyday speech. Through prosody, we bolster the messages carried by the words we speak—or create meanings directly opposed to them.

The sentence, "Those look great," is formed very differently by the vocal organs of a middle-aged man praising his friend's new khakis—and those of the khaki owner's teenaged son. One is a carefully articulated effusion of genuine praise, the other an artful act of deadpan sarcasm.

Irony and sarcasm are not the only way we use prosody. We use it every time we express tenderness, or anger, or enthusiasm, or any number of other nuanced emotional states that give the human voice its peculiar power to woo, persuade, threaten, cajole, and mollify. Prosody makes the difference between the affectless utterances of HAL the computer in *2001* (or Mr. Spock in *Star Trek*) and the rich and expressive instrument of Morgan Freeman or Meryl Streep—or even just the lilting, songlike way you say "*Hel*lo" when you answer the phone, so your caller doesn't think you're a machine. The term comes from the ancient Greek: *pro*, meaning "toward," and *sody*, meaning "song." We speak *toward song*. Except I didn't anymore, according to Zeitels.

"You're behaving through a veil of monotone," he went on. "When you talk, you can't express emotion properly. You can't change pitch, can't get loud, can't do the normal things that a voice does to express how you *feel*."

This hit me hard. I had not been consciously aware of these changes; but now that he pointed them out, I had to acknowledge that my range of expression had indeed diminished. I had, before developing a polyp, enjoyed exercising the emotive powers of my voice: as well as singing in a high school choir and in college coffeehouses (and, ill-advisedly, Jann's band), I had competed in public speaking contests, taking first prize in two poetry reading competitions at my high school ("Turning and turning in the widening gyre . . .") and winning a raconteur contest in college. Part of the fun of publishing my first book, which got me onto *Oprah* and a bunch of other TV and radio shows, was *talking* about the thing, exercising the public speaking skills that had lain dormant since that college competition. Though I could still drive my voice through the basic melodic shifts necessary to make my emotional state more or less known, it had become burdensome to do so (too much expressive talking still left me pretty wiped out at the end of a day); and my voice was by no means the precision instrument it had once been. More cudgel than scalpel, it would, when imbuing a word or syllable with special emphasis ("He said *what*?"), often break up, or cut out, altogether.

But that wasn't the worst of it. For Zeitels now added:

"*You* are not being transmitted by your voice."

That the voice is a vital clue to character and personality—to fundamental identity—was not news to me. We all make split-second judgments about people according to whether they speak in a deep, resonant, commanding baritone, or a high, piping soprano, or a girlish whisper. We draw inferences about everything from where they were born and raised (according to how they pronounce their vowels and consonants, their *accent*), to their socioeconomic status, to their degree of education. And, of course, people do this to us. When I moved to New York City from Toronto, at age thirty, I was regularly interrupted by strangers who would say, with a knowing smile, "What part of Canada are you from?" They had detected the distinctive way I pronounce the words "out" and "about"—Americans hear them as *oot* and *aboot*—which is a function of how I move my mouth when forming the sounds that Americans say as "ow" and I say as "oo"—a gesture of the tongue and lips that I learned in earliest infancy from hearing my (Canadian) parents and my kindergarten friends pronounce that speech sound and which was duly hardwired into the motor nerves that control my vocal organs when I talk. Because such aspects of voice are laid down during a critical period of brain development, "unlearning" them is extremely difficult—impossible for some. Today, after more than thirty years living in the United States, I have never lost my *oot* and *aboot*, and it never fails to make me feel a bit self-conscious about how my voice is telling strangers something intimate about me.

So, yes, I had always known that the voice is a kind of aural fingerprint, something unique to every individual and from which listeners draw strong inferences—hence my worry over sounding like a Bukowski-esque barfly or *Sopranos*-style heavy after my injury. But in "speaking around" that injury, I was apparently projecting a new personality into the world: a more monotone, less enthusiastic, less engaged personality.

But my polyp wasn't just changing how others perceived me; it was actually changing my *behavior*. "People with your type of injury withdraw from scenarios intuitively," Zeitels said. "It must be a nightmare in a loud New York restaurant."

It certainly was. Raising my voice above the din caused my vocal polyp to smash against my *healthy* vocal cord with extra force, creating swelling in both vocal cords that could take a week to subside and that made my rasp even worse. Loud restaurants, raucous parties, clubs, concerts—I tended to avoid them now, and when I did find myself in these environments, I deliberately clammed up. As a child, I had always been extroverted, verbal, *performative*—an aspect of my personality that everyone in my family attributed to my position in the birth order: I was the youngest of three boys all born within three years. Our sister came along four years after me. I thus occupied an ambiguous region: youngest *of the boys*, but not the youngest in the family. I must have seen it as a treacherous place to dwell (I *did* see it as a treacherous place to dwell), where it would be all too easy to be lost, eclipsed, overshadowed, forgotten. Accordingly, I learned early how to compete for attention, staking my claim on the airspace: I became the loudest, most verbal, and, I suspect, most irritating of the kids in our family. I was about four years old when, after improvising a stand-up routine in our kitchen ("Dad, if you're a 'doc,' why can't I dive off you?"), my parents announced that when

I grew up I would have my own talk show—"Just like Johnny Carson!" Music to my ears, and a further prod to seizing the floor, to raising my voice. None of this surprised Zeitels. "Oh yeah," he said, "you might have been brewing this polyp for decades before you sang in Jann's band. Wallflowers and introverts don't get this injury."

Feeling shaken, I said, "So—this changes my *life*, in a way?"

"Totally," he said.

In my article, I omitted all mention of my own voice injury, focusing on the microsurgeries Zeitels performed on his patients and his research into a gel-based filler—an "artificial vocal cord"—for repairing the kind of damage that Julie Andrews sustained from her botched operation. The article went over well, and it was suggested to me that I perhaps expand it into a book about "the voice, in general." My first impulse was to say "Not possible." As my own injury made clear, the voice is a deceptively simple-seeming subject (*you sing, you talk—big deal*) that actually touches on some of the deepest mysteries in the natural world: namely, how we communicate thoughts, emotions, personality, upbringing, and a lot of other personal data (including clues about race, mental health, social class, even sexual orientation), on tiny ripples of air that we beam into other people's brains by moving our lips and tongue while exhaling. An alien species watching us perform this bio-lingual-psycho-acoustical feat would no doubt think, "*This is unreal!*"

And it is. But how to get one's hands around so big and diffuse a subject? The difficulty in even saying what the voice *is* did not bode well for attempting a book. Is the voice singing? Talking? Is a *cough* voice? A laugh? "Indeed, it seems we know exactly what we mean by the word *voice* as long as we don't try to define it!" as Johan Sundberg, the world's foremost authority on the physiology of singing, put it in the introduction to his classic textbook *The Science of the Singing Voice* (1987).[2] Aristotle, who defined the voice as "the sound produced by a creature possessing a soul," explicitly ruled out coughing as voice because a cough does not call up a "mental image."[3] That is, words. Unfortunately, that definition also rules out the high clear sustained note that an opera tenor hits and which sends shivers through us despite the isolated vowel's calling up no specific "mental image" (especially if we don't understand Italian). To say nothing of the fact that, in the 1950s, a branch of speech science, called paralinguistics, emerged that convincingly showed that all manner of vocal noises (coughs, sighs, gasps, *ums* and *ers*) can be highly revealing of a person's inner state of mind and heart—and as such have a communicative salience that, even by Aristotle's definition, qualify them as voice.[4]

Add to these confusions the epistemological conundrum that the voice is, conceptually, impossible to "locate." It is "in" the speaker's body as an act of breathing and articulation, but doesn't exist until it is manifest in the air as a sound wave. Arguably, the voice comes into existence, *as* voice, only when someone is around to process that sound wave in the brain's auditory cortex. (In voice science, the answer to the philosophical riddle: "Does a tree that falls in a forest make a sound if there's no one to hear it?" is "No!") A final complication arises from the fact that what science calls The Voice—everything from the buzzing sound source in our throats, to the way we sculpt that buzz into speech sounds with movements of our mouths, to the rhythm and melody of spoken language or song—results from the synchronized actions of many distinct body parts (lungs, vocal cords, tongue, lips, soft palate, or velum), all of them originally designed (by natural selection) for quite different tasks. Which of these is the *voice*—some, all, none?

In short, it didn't seem possible to write a book about something that the smartest people in the world *couldn't agree on how to define.* Best to give up thinking about it.

Except I couldn't. Every morning before tackling whatever writing project was on the front burner, I would scribble free-associative notes on a legal pad, writing down whatever came to mind when I thought of the word "Voice." Those pages look like the jottings of a madman: "wooing," "weapon!!" "talking cure," "Ebonics," "stuttering? Lisp?" "Primal Scream," "Dylan," "baby talk—babbling," "opera," "HITLER," "transex," "code-switch," "Henry Higgins," "Rich Little?" "castrati?" and on and on. I did this for weeks. Gradually, some order emerged from the chaos, as certain words and ideas kept recurring, attaching themselves (with the swooping arrows I drew) to other words and ideas. I was reading widely: books on phonetics, on animal vocal communication, on human motor control (voice is a *physical* gesture, after all), on language acquisition in babies, on male and female voices. After a couple of months, a way of embracing all of the disparate topics within a single narrative began to coalesce.

The key, I realized, was to think about what makes the human voice different from that of every other creature. All mammals and birds use vocal noises to communicate vital needs, through an array of *oinks* and *squawks, chirps, barks* and *baahs*. Parrots can even expertly mimic human speech—but without any idea what they're saying. My wife and I have owned a succession of parakeets over the last thirty years and while every one of them learned to echo back to us a few phrases with varying degrees of clarity ("You're so cute!" "Oh, I love you!"), none was ever able to make the association between the word "seeds," upon which I patiently tutored them, and the food we gave them. Highly social and intelligent, our birds were perfectly capable of flying into the living room and squawking and chattering noisily to let us know we'd forgotten to fill the food cup. But none ever learned to save their energy by staying in the cage and simply saying: "Seeds." We are the only animal that can perform that miraculous feat: to make the link between a specific vocal *sound* and an object that exists in the world.[5] Which is to say, we alone have tamed all those barks and squawks and chirps and roars—domesticated them—into articulate speech.

I call it a "miraculous feat," but that understates the case considerably. It is the reason that we, as a species, rocketed to the top of the food chain. If you've read Yuval Noah Hariri's superb *Sapiens*,[6] you know that scientists usually attribute our ascent to *language*, a faculty that allows us to refer to events in the past or future, to allude to people and things not immediately present ("seeds!"), to elucidate abstract philosophical concepts, and to make complicated plans and goals that we share with others of our species. No other animal can come close to doing this. Birds, dogs, chimps, dolphins—you name it—use their voices to make in-the-now proclamations about immediate survival and reproductive concerns, including expressions of fear, anger, hunger, and mating urges. Our unique ability for language has thus been described as the great dividing line, the "unbridgeable Rubicon," between us and every other living creature. More than that, Hariri explains, it is the key to how we came to rule the earth, since it enabled early humans—a relatively slow-running, physically weak, easily-preyed-upon animal—to plan and cooperate and strategize with each other to outsmart bigger, faster, more lethal predators, to organize into groups (or tribes) of a greater size than any other animal (chimpanzees, our closest animal relation and the next closest in terms of cooperation, can manage about one hundred members per group), eventually to build the villages, towns, cities, and nations that have given us primacy over the planet and everything on it. Written language eventually speeded this process, but writing only came along about five thousand years ago, a blink of the eye in terms of human history. Up until then all verbal communication in our species was achieved via *speech*. So, I'm not disputing the grand claims for language made by Hariri and others. I just think we need to refine the concept, to emphasize that we owe our planetary dominion not to language alone, but to our special talent for turning that awesome attribute into sound. The voice.

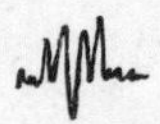

The current reluctance, in science, to accord the voice this special role in the life of our species seems to make sense. Language, after all, can be transmitted *without voice*. Deaf-mute people converse perfectly well using silent finger, hand, arm, and head movements, adding layers of emotional nuance to their "utterances"—*prosody*—through variations in body posture, speed of gesturing, and facial expressions. Writing is further evidence that voice is by no means necessary for language. This is clear from the very words I am typing and you are reading, words I'm building into grammatical structures that carry meaning, and upon which I'm also imposing a layer of prosody by strategically sprinkling commas, periods, dashes, and other punctuation—even resorting to *italics* and the odd exclamation point!

So, yes: the voice is by no means necessary for language.

Yet we would not have attained our present position at the top of the food chain had we been forced to rely on either sign language or writing alone to communicate. Gestural signing imposes severe limitations on sender and receiver, and in the brute struggle for survival that is evolution by natural selection, such things matter—a lot. The mute hunter-gatherer who spots a leopard a few yards off would, to notify the rest of his hunting party of this specific threat, have to turn and catch the attention of his scattered band and sign the word: "Leopard!"—thus risking being eaten before he dropped his spear to free up his hands. The hypothetical humans who relied purely on written language would be even worse off.

Imagine the confusions that might arise as the band huddled to read the lead hunter's hastily written missive. ("Does that say *leopard*?")

By then, it would be all over.

The human voice, in short, has a set of adaptive advantages that cannot be matched by signing or writing or indeed any other means of language transfer. The voice transmits words at a speed roughly five times faster than the movements of sign language. Lowered to a whisper, the voice can be heard in the pitch black by hunters or warriors stalking prey or enemies at night. The voice is unique in how it "splits" into as many channels as there are ears to hear it, so that a shout touches off the alarm in anyone within earshot (visual signs also "split," but only among receivers whose eyes happen to be turned toward the signaler).[7] Voice can travel great distances, penetrate dense jungle, travel around corners, and even through some solid barriers, yet leaves no trace—unlike footprints, scent, or other clues useful to a tracking predator. It can do all this even when the signaler is engaged in important tasks that occupy the hands, arms, and legs: nursing a baby; making a fire; fashioning a weapon; sewing clothing; raising the aforementioned spear—or running from an ambush by a group of invaders, the hands not gesturing or writing but bunched into fists and pumping hard. The shouted words, "Surprise attack—*run*!" might save a whole village.

For this reason, I challenge the current orthodoxy of language being the decisive factor in our species' rise. What ultimately put us on top is the faculty that makes language so potent—a faculty so ubiquitous, so everyday, yet so fleeting (the sound dying to silence, even as it leaves the lips), that we fail to remark upon it, but which would astonish that visiting alien species more than echolocation in bats or the intricate songs of the humpback whale.

Lest all of this smack of a certain tiresome human exceptionalism, or misplaced triumphalism (our stewardship of the planet might leave some things to be desired), let me quickly add that often the most useful information we convey with our voice is in those elements of the sound signal that are *not* language—not just the *timbre*, or "sound quality," like the rasp and rumble that made me sound like an underworld heavy, but the prosody, much of which was (as we will see) braided into our DNA over millions of years by our prehuman ancestors. It is upon *this* level of vocal signaling that a shaken employee draws when he gets off the phone with his boss and tells his coworkers, "Well, he *said* he liked the report, but I can tell he's going to fire me," or a wife who, upon hearing her husband ask for the remote, blurts out, "Are you having an *affair*?" (later telling a friend: "I just heard something in his *voice*").

The atavistic echo of nonhuman animal sounds (for marking territory and showing tribal kinship) are audible, as well, in the way we shape vowels and consonants—our regional accents (*aboot* which, more later). The dramatic timbral and pitch differences between men's and women's voices, along with certain changes in texture that occur in moments of sexual arousal, are evolved traits central to the continued existence of our species (through the erotic voice signals we send and receive and which spur our urge to mate). As we will see, how we pick our political leaders also depends more on primordial echoes of the beasts and predators from whom we evolved than might be immediately obvious, especially in times of political instability and division when those parts of our brain that respond to the voice's emotional channel are especially tuned to tones of fear and hate, anger and violence. In short, our collective fate as *Homo sapiens* (shaped, to a large degree, by the voices of our political leaders) relies far more on purely nonlanguage elements of speech than we might imagine or wish.

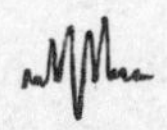

Likewise, our individual fates. Our career and romantic prospects, social status and reproductive success depend to an amazing degree on how we *sound*. This is a question not only of our vocal timbre, which is partly passed down by our parents (in the size, density, and viscosity of our vocal cords and the internal geometry of the resonance chambers of our neck and head), or our accent, but also our volume, pace, and vocal *attack*: elements of our speech that betray dispositions toward extroversion or introversion, confidence or shyness, aggression or passivity—aspects of temperament that are, science tells us, partly innate, but also a result of how we respond to life's challenges, in the innumerable environmental influences that mold personality and character and, consequently, our voice.

In listeners' ears, our voice *is* us, as instantly "identifying" as our face. Indeed, researchers in 2018 discovered that voices are processed in a part of the auditory cortex cabled directly to the brain region that recognizes facial features. Together, these linked brain areas make up a *person-differentiating system* highly valuable for ascertaining, in an instant, who we know and who's a stranger.[8] The voice recognition region can hold hundreds if not thousands of voices in long-term memory, which is why you can tell, within a syllable ("Hi . . ."), that it is your sister on the phone and not a telemarketer, and that Rich Little is attempting to "do" Bill Clinton and not Ronald Reagan (both of whose voices you can conjure in your auditory cortex as readily as you can call up their faces in your mind's eye).[9] That we do, sometimes, mistake family members for one another over the phone shows that not only are immutable anatomical attributes of voice (vocal cords and resonance chambers) as heritable as the facial features that make parent and child (or siblings) resemble each other, but that families often share a style of speaking, in terms of prosody, pace, and pronunciation. But the voice of every person is (like face or fingerprints) sufficiently unique, in its tiniest details, that such misidentifications are usually caught within seconds.

Indeed, it is a philosophical irony of cosmic proportions that the only voice on earth that we do *not* know is our own. This is because it reaches us, not solely through the air, but in vibrations that pass through the hard and soft tissues of our head and neck, and which create, in our auditory cortex, a sound completely different to what everyone else hears when we talk. The stark difference is clear the first time we listen to a recording of own voice. ("Is that *really* what I sound like? Turn it *off*!") The distaste with which so many of us greet the sound of our *actual* voice is not purely a matter of acoustics, I suspect. A recording disembodies the voice, holds it at a distance from us, so that we can hear with pitiless objectivity *all* aspects of how we speak, including the unconscious ways we manipulate prosody, pace, and pronunciation to create the voice we *wish* we had. When I mentioned this to a friend, he grimaced at the memory of hearing his recorded voice for the first time. "God!" he cried. "The *insincerity*!" He was reacting to the mismatch between who he knows himself (privately and inwardly) to be, and the person that he seeks to project into the world.

All of us do this, quite unconsciously, and until we hear ourselves on tape we remain mercifully deaf to how we perform this ideal self, in a bid to "put ourselves across," to make an impression. The enterprise of being human is to carve out a congenial place to occupy in the world, an achievement that we know, intuitively, to depend to a frightening extent on how our voices sound in the ears of others. This book isn't one of those instructional manuals that promises to give you a more *assertive* or *sexy* or *persuasive* voice—to aid you in the Darwinian struggle to advance in your job and land the partner of your dreams. But I hope that over the course of its eight chapters and coda it will have solved certain mysteries of the voice sufficiently to give you better insight than the *Fix Your Voice Fix Your Life!* come-ons that promise, through a few "easy breathing exercises," to transform you *overnight*! Something as negligible as a minuscule bump on one vocal cord changed who I am by altering my voice. It works in both directions. To alter your voice in ways that conform better to the person you feel yourself to be, or that you wish you were, means changing, fundamentally, who you are. It can be done, but not overnight.

In terms of structure, this book is a little like the vocal signal itself: it begins by examining how the voice is manifest in a single individual (a newborn baby), and then radiates outward like a sound wave, in ever-expanding concentric circles, from investigating that initial assertion of raw need (*feed me!*) to examine how we mold that cry into speech and then how we engage other voices, in the form of back-and-forth conversation between two people. The scope then widens, again, to look at how the voice works in the surrounding society: the voice as badge of tribal membership, status symbol, class marker, and racial identity, all factors that help to place us in the social matrix, and define us in terms of who we woo and win as romantic partner (straight, gay, lesbian, or trans—each of which has its own particular vocal signal). The outermost circles address the religious voice of exhortation and worship, and the public voice of mass broadcast (radio, television, movies) and, ultimately, the voice of political leadership, the single voices that steer our collective future. The voice of power does not always show our species at its best, but the singing voice invariably does, in my opinion, so I will spend a chapter exploring and celebrating that miracle. The book ends with a view of the aging voice and the wisdom that, if we are lucky, the old voice denotes.

Along the way, I'll touch upon the evolutionary pressures that created our uniquely human voice, its subtle emotional prosody and its game-changing specialization for language. Here, I depart from the prevailing view that our ability to shape our vocal signals into meaningful utterance results purely from changes to our brain that appeared some fifty- to sixty thousand years ago, spurring a massive spike in intelligence called the Great Leap Forward—a surge in cognitive power that supposedly caused language to somehow blip to life in our heads. Instead, this book emphasizes the role that *the voice itself* played in creating language in our species. That story stretches back much further than sixty thousand years—back, indeed, to when the first vertebrates emerged from sea onto land—and a story bolstered by recent research into the genetic mutations behind the blindingly fast and precise tongue and lip movements that enable speech.

We master this trick of high-speed articulation as infants, a feat of coordination between brain and body so amazing that some scientists have insisted that language must be innate. Newborns do indeed arrive in the world with a staggering amount of linguistic knowledge already present in the brain. But that knowledge derives, not from the fact that words, grammar, and syntax are necessarily preinstalled in us like the operating system on our computer, but because (recent science shows) our surprisingly long and intensive regime of voice-based training for language—the lessons we absorb through listening extremely closely to parents and caregivers—begins even before we leave the womb.

ONE

BABY TALK

The first experiments in fetal hearing were conducted in the early 1920s. German researchers placed a hand against a pregnant woman's belly and blasted a car horn close by. The fetus's startle movements established that, by around twenty-eight weeks' gestation, the fetus can detect sounds.[1] Since then, new technologies, including small waterproof microphones implanted in the womb, have dramatically increased our knowledge of the rich auditory environment[2] where the fetus receives its first lessons in how the human voice transmits language, feelings, mood, and personality.

The mother's voice is especially critical to this learning—a voice heard not only through airborne sound waves that penetrate the womb, but through bone conduction along her skeleton, so that her voice is *felt* as vibrations against the body. As the fetus's primary sensory input, the mother's voice makes a strong and indelible "first impression."

Monitors that measure fetal heart rate show that, by the third trimester, the fetus not only distinguishes its mother's voice from all other sounds, but is emotionally affected by it: her rousing tones kick up the fetal pulse; her soothing tones slow it.[3] Some researchers have proposed that the mother's voice thus attunes the developing nervous system in ways that predispose a person, in later life, toward anxiety or anger, calm or contentment.[4] Such prenatal "psychological" conditioning is unproven, but it is probably not a bad idea for expectant mothers to be conscious, in the final two months of pregnancy, that someone is eavesdropping on everything they say, and that what the listener hears might have lasting impact. The novelist Ian McEwan used this conceit in his 2016 novel, *Nutshell*, which retells Shakespeare's *Hamlet* from the point of view of a thirty-eight-week-old narrator-fetus who overhears a plot (though "pillow talk of deadly intent") between his adulterous mother and uncle.

As carefully researched as that novel is regarding the surprisingly acute audio-perceptual abilities of late-stage fetuses, McEwan takes considerable poetic license. For even if a fetus could understand language, the ability to *hear* speech in the womb is sharply limited. The uterine wall muffles voices, even the mother's, into an indistinct rumble that permits only the rises and falls of emotional prosody to penetrate—in the same way that you can tell through the wall you share with your neighbor that the people talking on the other side are happy, sad, or angry, but you can't hear what they're actually saying. Nevertheless, after two months of intense focus on the mother's vocal signal in the womb, a newborn emerges into the world clearly recognizing the mother's voice and showing a marked preference for it.[5] We know this thanks to an ingenious experiment invented in the early 1970s for exploring the newborn mind. Investigators placed a pressure-sensitive switch inside a feeding nipple hooked to a tape recorder. When the baby sucked, prerecorded sounds were broadcast from a speaker. Sounds that interested the infant prompted harder and longer sucking to keep the sound going and to raise its volume. Psychologist Anthony DeCasper used the device to show that three-day-olds will work harder, through sucking, to hear a recording of their own mother's voice over that of any other female.[6] The father's voice sparked no special interest in the newborn[7]—which, on acoustical grounds, isn't surprising. The male's lower pitch penetrates the uterine wall less effectively and his voice is also not borne along the maternal skeleton. Newborns thus lack the two months of enwombed exposure to dad's speech that creates such a special familiarity with, and "umbilical" connection to, mom's voice.

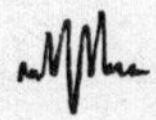

The sucking test has revealed another intriguing facet of the newborn's intense focus on adult voices. In 1971, Brown University psychologist Peter Eimas (who invented the test) showed that we are born with the ability to hear the tiny acoustic differences between highly similar speech sounds, like the *p* and *b* at the beginning of the words "pass" and "bass." Both are made by an identical lip pop gesture. They sound different only because, with *b*, we make the lip pop while vibrating our vocal cords—an amazingly well-coordinated act of split-second synchronization between lips and larynx that results in a "voiced" consonant. With the *p*, we pop the lips while holding the vocal cords in the open position, making it "unvoiced." We can do this with every consonant: *t*, voiced, becomes *d*; *k* becomes hard *g*; *f* becomes *v*; *ch* becomes *j*. Babies, Eimas showed, hear these distinctions at birth, sucking hard with excitement and interest when a speech sound with which they've become bored (*ga ga ga*) switches to a fascinating new one (*ka ka ka*).[8] Prior to Eimas's pioneering studies, it was believed that newborns only gradually learn these subtle phonemic differences.

The significance of this for the larger question of how we learn to talk emerged when Eimas tested if infants could discriminate between speech sounds from languages they had never heard—in the womb or anywhere else. For English babies this included Kikuya (an African language), Chinese, Japanese, French, and Spanish, all of which feature minuscule differences in shared speech sounds, according to the precise position of the tongue or lips, or the pitch of the voice. The experiments revealed that newborns can do something that adults cannot: detect the most subtle differences in sounds. Newborns, in short, emerge from the womb ready and willing to hear, and thus *learn*, any language—all seven thousand of them. This stands to reason, because a baby doesn't know if it is going to be born into a small French town, a hamlet in Sweden, a tribe in the Amazon, or New York City, and must be ready for any eventuality.[9] For this reason, neuroscientist Patricia Kuhl, a leading infant language researcher, calls babies "linguistic global citizens"[10] at birth.

But after a few months, babies lose the ability to hear speech sounds not relevant to their native tongue—which has huge implications for how infants sound when they start speaking. Japanese people provide a good example: when speaking English, adults routinely swap out the *r* and *l* sounds, saying "rake" for "lake," and vice versa. They do this because they cannot *hear* the difference between English *r* and *l*. But Japanese *newborns* can, as Eimas's sucking test shows. Change the *ra* sound to *la,* and Japanese babies register the difference with fanatic sucking.

But around seven months of age, they start having trouble telling the difference. At ten months old, they don't react at all when *ra* changes to *la.* They can't tell the difference anymore. English babies actually get better at it.

The reason is exposure and reinforcement. The ten-month-old English baby has spent almost a year hearing the English-speaking adults around her say words that are distinguished by clearly different *r* and *l* sounds. Not the Japanese baby, who spent the ten months after birth hearing a Japanese *r* that sounds almost identical to our English *l*, the tongue lightly pushing against the gum ridge behind the upper front teeth. Because there is no clear acoustic difference between the Japanese *r* and the English *l*, Japanese babies stop hearing a difference. They don't *need* to, because their language doesn't depend on it.

All of which is to say that the developing brain works on a "use it or lose it" basis. Circuitry not activated by environmental stimuli (mom's and dad's voices) is pruned away. The opposite happens for brain circuits that are repeatedly stimulated by the human voice. They grow stronger, more efficient. This is the result of an actual physical process: the stimulated circuits grow a layer of fatty cells, called myelin, along their axons, the spidery branches that extend from the cell's nucleus to communicate with other cells. Like the insulation on a copper wire, this myelin sheath speeds the electrical impulses that flash along the nerve branches that connect the neurons which represent specific speech sounds. Neuroscientists have a saying: "Neurons that fire together, wire together"—which is why the English babies in Eimas's experiments got *better* at hearing the difference between *ra* and *la*: the neuron assemblies for those sounds fired a whole lot and wired themselves together. Not so for Japanese babies.

In short, the voices we hear around us in infancy physically sculpt our brain, pruning away unneeded circuits, strengthening the necessary ones, specializing the brain for perceiving (and eventually producing) the specific sounds of our native tongue.

Some infants fail to "wire in" the circuits necessary for discriminating highly similar sounds. Take the syllables *ba, da,* and *ga*, which are distinguished by where, in the mouth, the initial sound is produced (*b* with a pop of the lips; *d* with a tongue tap at the gum ridge; *g* with the back of the tongue hitting the soft palate, also called the velum). These articulatory targets determine how the resulting burst of noise transitions into the orderly, musical overtones of the *a*-vowel that follows: a sweep of rapidly changing frequencies, over tens of milliseconds, that the normal baby brain, with repetition, wires in through myelinating the correct nerve pathways.

But a small percentage of babies, for unknown reasons, fail to develop the circuits for detecting those fast frequency sweeps. Sometimes a child registers *ba*, sometimes *ga* or *da*. Parents are unaware of the problem because kids compensate by using contextual clues. They know that mom is saying "bat" and not "pat" because she's holding a bat in her hand. They know dad is talking about a "car" because he's pointing at one. The problem surfaces only when the child starts school and tries to learn to read. That is, translate written letter-symbols into the speech sounds they represent. He can't do it, because his brain hasn't wired-in the sounds clearly. He might read the word "dad" as "bad," or "gab," or "dab." These children are diagnosed with dyslexia, a reading disorder long believed to be a vision problem (it was once called "word blindness"). Thanks to pioneering research in the early 1990s by neuroscientist Paula Tallal at Rutgers University, we now know that some dyslexia is a problem of *hearing*, of processing human voice sounds.[11] Tallal has been helping to devise software that slows the frequency sweeps in those consonant-vowel transitions so that very young children can train their auditory circuits to detect the different speech sounds, and thus wire them in through myelination of the nerve pathways. All to improve their *reading*.

Of course, to learn a language, it is not enough simply to recognize the difference between *pa* and *ba,* or *la* and *ra.* To understand speech—and to produce it one day—babies must accomplish another exceedingly difficult feat of voice perception. Though it might seem, to us, as if we insert tiny gaps of silence between words when we speak (like the spaces between words on a printed page), that's a perceptual illusion. All voiced language is actually an unbroken ribbon of sounds all slurring together. To learn our native tongue, we had to first cut that continuous ribbon into individual words—not easy when you're a newborn and have no idea what any words mean. You can get an idea of what you were up against by listening to a YouTube clip of someone speaking a language you don't know: Croatian, or Swahili, or Tagalog. Try listing ten words. You can't do it because you can't tell where one word ends and another begins. This is the problem you faced at birth—and, by around eight months, had solved.

Here's how. Despite appearances, babies, reclining in their strollers or lying in their cribs, are anything but passive receptors of the speech that resounds all around them. Indeed, even before birth—from the seventh month of gestation onward—the fetus runs a complex statistical analysis on the voices it perceives, and registers patterns. The sucking test shows that one pattern newborns detect is word stress.[12] English, on average, emphasizes the first syllable of words: *con*tact, *foot*ball, *he*ro, *sen*-tence, *mom*my, *pur*ple, *pig*eon; words that emphasize the second syllable (like sur*prise*) are far less common. In French, it's the reverse—a weak-strong pattern: "bon*jour*," "mer*ci*," "vi*tale*," "heu*reux*." Babies zero in on these patterns and use them to locate word boundaries. Take a mystifying sequence of speech sounds like:

staytleeplumpbukmulaginkaymfrumtheestarehed

An American baby will apply English's strong-weak probability to identify the first sound clusters (*staytlee*) as a possible stand-alone word (STAYT-lee—or "Stately"). The next two syllables, however (*plumpbuk*), don't make an English word, no matter what stress pattern you apply (*PLUMP*-buk; plump-*BUK*). To deal with that, the baby uses another type of statistical analysis. In all languages, the likelihood that one speech sound will follow another is highest *within* words, less likely across words. Patricia Kuhl supplies a good example from Polish, where the *zb* combination is common, as in the name Zbigniew.[13] But in English *zb* occurs only across word boundaries, as in "leaveZ Blow" or "windowZ Break"—and thus crops up less frequently. Sophisticated listening tests show that eight-month-olds use these "transition probabilities" to segment the sound stream—and babies can do this after *just two minutes'* exposure to a stream of unfamiliar speech sounds.

This staggering speed of learning speaks to Darwin's assertion, in *The Descent of Man*, that speech acquisition in children reveals not an instinct for *language*, but an instinct to *learn*—as in an English baby's lightning-fast realization that the *pb* in *plumpbuk* is illegal and that it makes sense to split the speech stream there, to create the separate chunks *plump* and *buk*. Eventually, the child will use both statistical strategies to help segment the entire sequence and arrive at the first words of James Joyce's *Ulysses:*

Stately, plump Buck Mulligan came from the stairhead . . .

She will accomplish this stunning feat before her first birthday, well before she has the least clue about what any of the words actually *mean*. But in snipping the sound ribbon into its separate parts, the baby stands a chance of figuring out how to assign meaning to each small cluster of sounds—clusters we call "words."

Babies do not do all this work on their own. They receive significant help from adults, who unconsciously adopt a highly artificial vocal style when addressing them.

Remarkably, no language expert took any formal notice of the unusual way we talk to infants until 1964, when Charles A. Ferguson, a linguist at Stanford University, published the paper "Baby Talk in Six Languages." It catalogued the identical way parents speak to babies in a slew of widely different tongues, including Syrian Arabic, Marathi (a language of western India), and Gilyak (spoken in Outer Manchuria), as well as English and Spanish. In each instance, caregivers prune consonants (as when English parents use "tummy" rather than "stomach") and use onomatopoeia (in English, "choo choo" for "train," and "bow wow" for "dog").[14] Ferguson was not, however, investigating how babies learn to speak—you could even say he was doing the exact opposite. He was searching for evidence to support the theory, first advanced by linguist

Noam Chomsky, that language is not learned at all, but is instead inborn, preinstalled in the brain before birth.

Chomsky made a strong argument for this when he pointed out, in a series of now famous books and papers of the late 1950s and early '60s,[15] that parents don't sit around systematically teaching newborns how to talk. Instead babies acquire speech from hearing only the half-mumbled, sporadic, often ungrammatical talk all around them (like the murky, overlapping conversations in a Robert Altman movie). Despite this "poverty of the stimulus," as Chomsky called it, children by age four speak in complex multi-word sentences: forming questions from statements, embedding clauses, speaking in past and future tenses. They can accomplish this, Chomsky said, only because language *already exists in the brain*. "In fact," he once said, "language development really ought to be called language growth because the language organ grows like any other body organ."[16] This figurative "language organ" cannot be dissected like a liver or heart, Chomsky said, but it can be *described* through analysis of the syntax common to all languages—the "deep structures" that make up what Chomsky called "Universal Grammar," the innate rules that govern all languages (no matter how different they sound on the surface).

Ferguson, in studying the baby talk of such vastly different cultures, was in search of Chomsky-inspired "language universals." The first study to address how adult "baby talk" helps infants *acquire* language did not appear until 1971, when Catherine E. Snow, a twenty-six-year-old graduate student at McGill University, stumbled onto the topic by chance.

Like Ferguson (and most social scientists of the day), Snow accepted Chomsky's claim that language is innate, so when she was invited to lead a graduate seminar on language acquisition, she planned to do so from the Chomskyan perspective. In the interest of thoroughness, however, she decided to look up the evidence upon which Chomsky based his claim that infants hear mostly garbled, incomplete, stuttered, overlapping, highly degraded speech—his primary proof that language must be inborn. Snow discovered that no papers existed supporting his "poverty of the stimulus" argument. Chomsky had apparently relied on his subjective impression of what babies must hear. Snow was amazed, and aghast, later saying: "I felt somehow offended that linguists made, accepted, and uncritically propagated claims about such matters with no sense of obligation to make the relevant observations."[17]

For her PhD dissertation, Snow designed lab studies to learn what babies actually hear from caregivers. She recorded thirty women (some mothers, some not) talking to children, of various ages, on set topics. The recordings revealed that the mothers spoke entirely differently to children of one month to two years old than to children who are ten years old, and that the childless women also adopted these unique features of infant-directed speech. All the women used the simplifications Ferguson had documented (pared-back sound clusters, onomatopoeic words) and also heavy repetition of new words. "Put the red truck in the box now," one mother told her two-year-old. "The red truck. No, the red truck. In the box. The red truck in the box." Such systematic redundancies helped infants segment the speech stream (the sounds "red" and "truck" and "box" jump out), while the short, simple utterances, each containing just one idea, helped babies detect how sentences are constructed. "That's a lion," one mother told a toddler. "And the lion's name is Leo. Leo lives in a big house. Leo goes for a walk every morning. And he always takes his cane along." Snow concluded that, contrary to Chomsky's claim, infants "do not learn language on the basis of a confusing corpus full of mistakes, garbles, and complexities. They hear, in fact, a relatively consistent, organized, simplified, and redundant set of utterances which in many ways seem quite well designed as a set of 'language lessons.'"[18]

Snow's findings didn't mean that our language capability is not to some degree inborn; clearly, we possess a biological, genetically determined capacity for speech—otherwise, we wouldn't be able to do it. But Snow's work provided a good and necessary corrective to the pendulum swing that had made language, under Chomsky's model, seem entirely a result of "nature," with no role played by "nurture"—no role played by the human voice.

Snow's findings, published in 1972, sparked an explosion of follow-up studies challenging Chomsky's view. In 1977, Olga Garnica, an assistant professor at Ohio State University, published a groundbreaking paper focused on the artificial, exaggerated prosody that caregivers automatically adopt when speaking to infants and toddlers—the high-pitched, slowed-down, singsong speech familiar to anyone who has ever heard someone talk to a baby ("Nowww . . . aren't you . . . *Keeeyyy*-OOOOT?").[19] Pitch peaks on specific words, extended pauses between words, and long-drawn vowels are, Garnica said, all part of a system to help babies segment the speech stream, hear how grammar works, and detect the specific tongue and lip positions that distinguish an *ee* from an *oo*, or an *ah* from an *uh*. Stanford University linguist Anne Fernald showed that these prosodic exaggerations are adopted by parents in all cultures and languages and that every adult uses them when talking to babies (whether they're aware of it or not).[20] Furthermore, children as young as four adopt the infant-directed singsong when talking to two-year-olds—or to their dolls. We do it when speaking to our *pets* and when talking to foreign-accented strangers who ask us for directions in the street, strong evidence that high-pitched, singsong, slowed-down speech is an adaptive vocal mechanism that evolved in our species for teaching language.

Physiologically, this speaking style makes sense: babies are best at detecting high-pitched sounds (not until ten years old do they acquire the low-frequency hearing typical of adults[21]). The artificially high pitch grabs and holds the infant's attention and, when used along with the simplifications and repetitions described by Ferguson and Snow, it becomes part of an elaborate voice-based language-tutoring system that linguists call Motherese. Remarkably, Motherese works, Fernald showed, in a feedback loop between parent and baby: as the infant starts to speak, forming first syllables and words, the adult's Motherese automatically adjusts itself, the raised pitch progressively lowering, the word simplifications and repetitions gradually diminishing, in inverse proportion to the child's mastery of its native tongue.

Most babies are a year old before they use all the linguistic information they've been hoarding since the womb—and utter their first word. But they don't, to put it mildly, spend those first twelve months in vocal silence.

The first act that every healthy baby commits upon emerging from the womb is a cry—an expertly coordinated spasm of the diaphragm, with an exquisitely timed closure of the vocal cords across the windpipe (so that they vibrate and produce sound, an act linguists call "phonation") and a synchronized opening of the mouth and lowering of the tongue: *Waaaahhh!*

That newborns can, without a single rehearsal, perform this act of complex physical coordination upon first exposure to the air (before birth, all humans are aquatic animals) suggests that the infant cry is pure instinct, like the reflex kick of your foot when the doctor taps your knee with his hammer. And it has a clear, biological survival purpose: it ejects from the windpipe any mucous or amniotic fluid on which the baby could choke. But it also has a vital function as *communication*. It notifies everyone within earshot that the screamer is alive.

In the first days and weeks of life, the baby's cry grows more robust as the abdominal muscles and diaphragm strengthen with use, and as the baby gains greater control of its tongue and lips, instinctively shaping the resonance chambers of throat, mouth, and lips to boost the signal to a window-rattling volume (people who study to be opera singers have to relearn how to do what a baby does naturally, as we'll see later). This sonic blast gives the otherwise helpless creature the ability to summon, from a great distance, its mother. Consequently, the infant cry has been called an "acoustical umbilical cord."[22]

It has also been called a "biological siren," and like any siren, it was engineered (by nature) to be intensely annoying.[23] A typical baby's cry has a fundamental frequency (or pitch) around 500 cycles per second (five times that of an adult male voice) with overtones (that is, the additional audible pitches that are part of every complex vocal sound) around 1,400 and 5,700 cycles per second—very high frequencies that overload the human auditory cortex. Like nails scraping a blackboard, or the rattle of a jackhammer, the cry causes great psychological distress in those who hear it, so they *must* spring into action and tend to the baby's needs, if only to alleviate the assault to their *own* nervous system. Thus, the paradox in the baby's cry, as described by Debra Zeifman, a psychologist at Vassar College who specializes in mother-infant bonding: "part of [the cry's] power to activate caregiving lies in its noxiousness, and . . . this very noxiousness can also evoke abusive or avoidant responses by caregivers."[24] (Parents convicted of injuring, or even killing, infants with shaken baby syndrome frequently offer as defense that, "She wouldn't stop *crying*.")

In the late 1950s, psychiatrist Peter Ostwald, of the University of California School of Medicine, became fascinated by the baby's cry and its uncanny similarity to the vocal acoustics of severely ill psychiatric patients. In a 1961 paper in the *Archives of General Psychiatry*, Ostwald isolated the universal "stress tone" in patients suffering from acute depression, schizophrenia, and psychoneurotic hypochondria.[25] The voices of all these patients showed anomalies in the higher overtones centered around 500 cycles per second—the average pitch of the baby's cry. In hysterical patients, the overtones spiked, giving the voice a "sharp" tone of complaint; in obsessional depressives, the overtones were level, giving the voice a "flat" and "irritating" quality; in brain-damaged patients, the overtones dropped in pitch across utterances, giving the voice a "hollow" and "emotionally drained" sound; in grandiose patients, the overtones rose and stayed level, resulting in a voice "characteristic of persons who spoke loudly, emphatically, and needed to be heard, to impress, and to influence others." Ostwald saw this as acoustic confirmation of Freud's theory that emotional disorders reflect an ur-injury from earliest childhood, a psychic wound that regresses the adult sufferer to a state of infantile need and complaint that can actually be *heard* in the pitch and timbre of the sufferer's voice.

The connection between the baby's cry and the sound of clinical neurosis would, less than ten years after Ostwald's study, inform a treatment pioneered by a California-based psychotherapist named Arthur Janov. During a group therapy session at Janov's San Francisco clinic in 1967, an overwrought twenty-two-year-old male patient fell writhing to the floor and began to emit what Janov later described as "an eerie scream welling up from the depths"—a sound that "one might hear from a person about to be murdered."[26] The screaming fit, reportedly, alleviated the patient's neurosis. Janov began encouraging his other patients to scream. They, too, felt better afterward. Janov called the noise the "Primal Scream" and said that it was a more effective treatment for neurosis than psychotherapy or drugs. The scream, he said, regresses patients to a period before certain emotional injuries were inflicted, injuries that create a permanent "muscle tension" throughout the body (an idea he borrowed from the influential psychologist, and student of Freud, Wilhelm Reich). This stored-up "psychic pain," Janov said, leads to a "clamping" of the respiratory and vocal muscles that is heard in the "squeezed" voice of the neurotic.[27] Janov claimed that the violent muscular spasms involved in screaming unlock this muscle tension and relieve the psychic pain, permanently.

When my neighbor Andrea teaches Kristin Linklater's "Freeing the Natural Voice" technique, she sometimes encourages loud, uninhibited vocal noises that release muscle tension, although not for psychotherapeutic purposes, but rather to free up the voice's acoustic range and power. Nevertheless, a friend of mine who actually traveled to Orkney, Scotland, to take a weeklong workshop withLinklater herself (who at eighty-three years old was still teaching her vocal technique to those willing to make the pilgrimage) reported that the participants, when encouraged to roll around on the floor and adopt various unusual postures while screaming, manifested a striking psychological side-effect. Every person in the workshop, my friend told me, at one point or another, "cried about their mother."[28]

While anxiety and other emotional disorders do affect the voice—tensing the respiratory muscles, which dampens volume; tightening laryngeal muscles and causing the voice to tremble; freezing muscles of the face and tongue, blurring articulation—there's no evidence that severe neurosis can be permanently alleviated by screaming, primal or otherwise. Today, any reported effectiveness in Janov's screaming treatment is understood to be a placebo effect, or short-term emotional catharsis.[29] Nevertheless, Janov's book *The Primal Scream*, published in 1970, sold millions and attracted celebrity adherents, including the actor James Earl Jones and Apple founder Steve Jobs. But the most famous proselytizer for Primal Scream Therapy was John Lennon. He underwent the treatment after Janov sent an unsolicited prepublication copy to Lennon's estate in England. The singer's openness to trying new therapies, religions, and drugs was well known,[30] as was his psychic turmoil: abandoned by both parents at age four and raised by an aunt, he was only just growing close to his mother, in his teens, when she was killed by a student driver; to this trauma was added, shortly before he read *The Primal Scream*, his split from the Beatles, divorce from his first wife, and a descent into heroin addiction.[31]

Janov personally oversaw Lennon's treatment, which lasted five months. Shortly afterward, Lennon released his first post-Beatle LP, *John Lennon/Plastic Ono Band*, which features songs inspired by Primal Scream, including the harrowing "Mother," where Lennon repeatedly shrieks "Mommy come *hoooooooome*," his voice growing more desperate, more atonal and ragged, with each iteration of "home." The effect is genuinely spooky: his final long-drawn cry, on the fade out, could be a newborn wailing in vain for its mother. In interviews, Lennon praised Primal Scream ("You're so astounded by what you find out about yourself")[32] but a few years later he relegated the treatment, and Janov, to the heap of cast-off therapies, religions, drugs, and gurus he had embraced and abandoned since the mid-1960s.

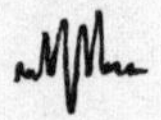

Although automatic, the newborn's "biological siren" also carries an echo of the child's native tongue, a linguistic palimpsest imprinted on the fetal nervous system during the final two months of listening in the womb. When researchers compared the cries of newborns from France and from Germany,[33] they discovered that the French two-day olds wailed on a rising pitch contour, mirroring the melodic pattern of spoken French; German newborns cry on a downward arc typical of that language's prosody.

The study's authors saw this astonishingly early mimicry of the maternal voice as a crucial adaptation to attract the mother's attention and "foster bonding."

But forming actual words, in any language, is anatomically impossible for all newborns and remains so for many months. This is owing to the extraordinary fact that we emerge from the womb with a larynx in the same high throat position as that of adult chimpanzees—which is to say that it is not located around the middle of the neck, as in adult humans (the bulge of the male Adam's apple is actually the pointy cartilage at the front of the larynx, to the inside surface of which one end of the vocal cords attach). Instead, the newborn's larynx is crowded up into the back of the mouth, close to the opening of the velum. This aids breastfeeding by creating an uninterrupted airway from nose to lungs (so newborns can suck at mom's breast without having to stop and "come up for air" as they feed—the milk flowing around the sides of the raised larynx and into the stomach).

But this high larynx position also severely restricts the range of vowels that any newborn can utter. And the ability to produce clear, distinctly different vowels, one from another (for instance, *ee, ahh, ooo*), is crucial for articulate speech; it's how you make, for instance, the separate words *had, heed, head, hide, hid, hood, who'd, Hud* from the same set of consonants; or *dad, dead, deed, did, Dodd, dud.* By altering slightly the curve and position of the tongue in the oral cavity, you change the relative size and shape of the different sections of your vocal tract, which runs from your vocal cords vertically up your throat and, after a ninety-degree bend, into the horizontal section of your mouth. Though they form a continuous tube, the vocal tract's throat and mouth sections act as two independent resonance chambers—and you boost certain vowel-defining overtones in the vocal spectrum depending on how you shape those resonance chambers with your tongue. For the *ee* sound, you lift the tongue toward your palate and push it forward to make the mouth resonance chamber small, which boosts the higher-pitched overtones (like a small-bodied violin); but by pushing your whole tongue forward, you simultaneously *enlarge* the throat resonator, boosting the lower-pitched overtones (like a big-bodied cello). The *blended* pitches produce the complex sound we hear as *ee*. The reverse happens with *Ahhh*, when you drop the back of your tongue, making a big mouth resonator and a small throat resonator. The lips get into the act when you make the *ooo* and *oh* vowels, rounding and extending the lips, which lengthens the entire vocal tract and lowers the pitch of *all* the overtones in the voice spectrum—in exactly the same way that a trombonist makes low notes by pushing out the slide on his instrument, lengthening the resonance tube.

The tiniest changes to the size or shape of the vocal tract's resonators has a huge effect on the different sounds our brains perceive, which is how English speakers, through subtle adjustments to the tongue and lips, produce the twenty-odd vowels of English, or Swedish speakers make the *forty* distinct vowels of their language. Even the slight pulling in of the lips against the teeth when we smile shortens the vocal tract enough to raise the signal's entire overtone spectrum, "brightening" the sound so that we can tell, over the phone, that the person speaking to us is in a good mood (you *hear* a smile). You also detect a sulky mood in the sound of a pout, which extends the lips, lowering the overtone spectrum. (Which is why photographers, when they want you to assume the expression of a happy person, shout: "Say cheese!" and not "Say choose!")

Now, imagine that you didn't *have* a throat resonator because (like a baby or a chimp) your larynx is pushed up into the back of your mouth. You'd be limited to the vowels that can be made only with the mouth resonator—a sound linguists call the schwa. Kind of a short *e* sound, it's actually the most common speech sound in all languages, as well as being the sound you make when pausing for thought ("uhhh"). Call it the sound of the cerebral cortex in neutral gear. It's useful in its place (at the end of words like "the"), but not so good if it were the only vowel sound we could make. The sentence "Who hid the head in the hut—Hud?" would come out as "Huh hud thuh huhd uhn thuh huht—Huhd?" If so vocally challenged a species managed to zoom to the top of the food chain, it would emphatically *not* be because of its ability to articulate clear, unambiguous vowels.

And it's why a newborn is *physically* incapable of producing any human language. Only as the baby transitions from liquid to solid food does the larynx descend down the throat, literally inching down the neck, day by day. As it does so, the larynx pulls the root of the tongue down with it (the back of our tongue is attached to the larynx by a system of ligaments). This elongation of the tongue down our throat is crucial to speech, because it is the tongue's vertical section that we manipulate (pushing it forward and backward) to produce the correct throat overtones for clear, well-articulated vowels.

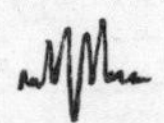

As the baby's larynx descends down the throat in the first months of life, she also gains considerable motor control of her articulators and starts to make an array of speech-like lip-pops for *p* and *b*, percussive tongue hits for *d* and *t*, fricatives and sibilants (like *s* and *sh*, which break the sound wave up into a hissing turbulence by pushing it through a narrow gap between tongue and teeth), as well as nasals, like *m* and *n*, by opening the velum and sending the soundwave through the nose. But not until she is six to eight years old will her larynx descend to the point where she can make vowels as finely sculpted as an adult's,[36] but even by her first birthday her larynx will have descended enough so that adults can infer what vowel sounds she's trying to make—a good thing, since it is at precisely this moment in infancy, at one year old, that she will put the various voice sounds together to utter her first word.

A baby's ability to translate its mental knowledge of language into spoken utterance is an extraordinary and, until recently, mystifying accomplishment. Babies do glean some important lessons in the complicated gymnastics of speech from peering at a speaker's mouth (my own home videos of our newborn son show him staring with rapt fixity at the movement of my own lips, as I natter at him from behind the camera, and he can even be seen clumsily trying to mimic the moves). But such "lip reading" can tell a baby only so much. It cannot, for instance, show a baby when and how to snap the vocal cords closed across the windpipe to make the voiced *b* that is so different from the unvoiced *p*, or what inner mouth target to hit with the tongue to make a *t* or a *g* sound, or exactly how big a gap to leave between the tongue and the back of the teeth to go *ssss*. And yet, around one year of age, most babies use these stunningly well-calibrated, precise, essentially invisible maneuvers to utter their first word.

They learn to do this the same way they learn everything: play. Tireless, dedicated, focused, trial-and-error play. Specifically, *vocal* play, which speech experts call "babbling"—an activity that begins, around four months of age, with utterances like *ba, ba, ba* or *da da da* or *ga ga ga*. This repetition of single sounds is called reduplicated babbling and it morphs, around eight months, into variegated babbling, which features the trickier task of mixing up various sounds, both voiced and unvoiced, in word- and sentence-like strings: as in *kaga-bodee* or *paba-tee-no*. These vocalizations were long seen as simply a way for the baby to strengthen the articulatory muscles. Today, experts in child development understand babbling to be the single most important stage in speech acquisition. Without it, we would never be able to tune our voice to the sounds of spoken language.

In babbling, babies listen closely to the sounds that emerge from their own mouth, and they compare these to the speech sounds of their native tongue—all that linguistic information they've been busily storing up in their auditory cortex since their seventh month in the womb. When, through random, playful movements of lips, tongue, velum, and larynx they get an accidental "hit"—a match between a stored speech sound and the noise they make with their voice—they get excited and repeat the sound (*babababa* or *mamama* or *dadada*), etching the instructions for these gestures into a part of the brain that is responsible for learning, and then coordinating and sequencing, highly complicated bodily movements: an ancient set of structures deep in the brain called the basal ganglia.[37] It's the same part of the brain you use when you learn to ride a bike or throw a ball (or, indeed, learn to walk, at around age one). At first, these actions are ill-coordinated, clumsy, until, through relentless practice, the basal ganglia get the moves sorted out and etched so deeply in your muscle memory that making an unvoiced *pa* versus a voiced *ba*, or the *l* sound in *la* versus the *t* in *ta*, seems unlearned, automatic. Not surprisingly, researchers have been investigating abnormalities in the basal ganglia, and its connections to the brain's "language centers," as a primary cause of stuttering.[38] But, as with all aspects of voice, "mechanistic" explanations are only part of the story. Psychology, acting in concert with biological causes, also plays a role, as John Updike suggests when exploring the roots of his own stammer in his memoir, *Self-Consciousness*. He stuttered when he felt himself to be "in a false position," as when addressing his high school as class president: "I did not, at heart, feel I deserved to be class president . . . and in protest . . . my vocal apparatus betrayed me."[39]

Birds undergo the same process of vocal learning as us when they acquire their species-specific songs in infancy. Ornithologists call birds' early subsong "babbling," and like human speech it involves sentence-like strings of sounds that the baby bird makes in imitation of its adult tutors, grooving into its version of the basal ganglia the proper sequence of tongue and beak movements that turn the phonation formed by their version of the larynx (it's called the syrinx and is located deep in the bird's chest) into the stream of twittering birdsong unique to a particular species. Which is why parakeets and other parrots can learn to mimic human speech—and it's also why, if you move a newborn bird to a community with a subtly *different* song than that of its birth parents, the displaced bird's voice will (as Darwin noted) take on the pitches, syllable lengths, and rhythms of its adoptive community, a "provincial dialect"—an *accent*.

The same of course happens to us. A human baby born into an affluent neighborhood of London will, during the babbling stage, wire in the motor circuitry for making an *aw* vowel sound in words like "dance" (*dawnce*), because that's what it hears from parents and others. But transport that same baby to New York during the babbling stage and it will wire up speech circuits for saying "dance" with an *ah* vowel sound. Parisian babies use their babbling to build motor pathways for easily articulating a sound like *eu*, in the word *neu*, which requires a high-front-tongue position (as in *ee*), with the lips, not pulled against the teeth as in English, but extended in a pout (as if saying *oo*). For Americans or Brits, who don't hear this *eu* sound as babies, producing a proper French *eu* feels as awkward as trying to rub your stomach and pat your head at the same time; for French kids, who practice the oral gestures for a year, grooving them into the basal ganglia, it's second nature.

Consonants are also etched into the basal ganglia during babbling: babies raised in India learn to curl the tongue back to tap the tip against the high part of the palate for *t* and *d* to match the sound their parents make, whereas babies in Texas tap the tip against the gum ridge, and French babies mash the body of the tongue against the back of the teeth, like Catherine Deneuve saying *tu*.

An important speech variable that linguists call "Voice Onset Time" is also learned in the babbling stage: the neural instructions for when, precisely, to turn on the vibration of the vocal cords to transform a *p* into a *b* or a *t* into a *d*. So sensitive to the timing of voice sounds are our ears that we can actually hear shades of ambiguity *within* the 65 millisecond VOT difference that distinguishes saying *pa* from saying *ba*. Hindi speakers, for instance, when saying *pa*, start phonating the *a* about 21 milliseconds sooner than do English speakers, giving a little bit of overlapped voicing to the initial lip pop.[40] You can hear that tiny difference. I was recently listening to CNN broadcaster Fareed Zakaria, whose native language is Hindi, talking about Islam and the "path to reform," except that it sounded like he was ever-so-slightly saying "*bath* to reform." He was turning his vocal cords on a shade earlier than a native English speaker. VOT also explains something that has obsessed me since I first saw the movie *Wait Until Dark* at age ten and noticed that Audrey Hepburn says words like "can't" and "call" a little bit like she's saying *gan't* and *gall*. I always assumed that this was some kind of finishing school affectation, since her overall accent is upper-class British. Then I learned that she was born and raised in Brussels. Her original language, Dutch, has a slightly faster voice-onset time for *a* after hard *c* than does English. So, when Hepburn says *gan't* for *can't* you are hearing not an affectation, but an inexpungable aspect of her earliest childhood—which is what you hear when most people speak, unless they've taken pains to remove such clues from their voice. That is, by changing their accent.

Even if some subtle markers like Voice Onset Time persist, the accent you learn in babyhood from your parents *can* be unlearned. In fact, children who are the offspring of foreign-accented parents don't even have to undergo formal training to lose their parents' pronunciation—it happens automatically if they start early enough. And most children do, when they leave the linguistic bath of home and go to school at age five. Henry, the eldest son of my Australian friends, Tony and Leslie, spoke with his parents' strong *thray-anitha-shrimp-ahn-thah-bahbie* Aussie twang, until he started attending a New York City elementary school. By Easter, his "Strine" accent was gone and his speech was indistinguishable from that of a native New Yorker. I recall, with some guilt, how my *own* son came home from first grade at his Manhattan public school in tears because his classmates had mocked him for saying "Sorry" with the low-back "o" vowel he had imprinted from his immigrant Canadian parents. Soon, and with no conscious effort, he rid himself of our mortifying "o," replacing it with a "proper" American version (which sounds, to my Canadian ears, like "Saaarry")—just one example of how the primary influence on human behavior shifts from parents to peers when children enter the Darwinian arena of the schoolyard. The voice, like every other behavior, adapts for survival.

But only within a certain time window, or "critical period."

Critical periods are developmental stages in childhood during which particular skills must be learned, or they will fail to develop altogether and after which they are locked in, more or less permanently. That speech and accent acquisition are subject to critical periods was elegantly demonstrated in studies of very young children who had suffered brain damage in an important language-processing region called Broca's area—a half-dollar-sized patch of cells on the surface of the left hemisphere, near the temple. Activity in Broca's area is how we construct our mental sentences before we say them: slotting the correct speech sounds into words, like a Scrabble player shifting letter tiles on her rack, and arranging those words into the right order. Broca's area then passes this information to the part of the brain that initiates the movements of the lungs, larynx, tongue and lips that turns a thought into sound. People with damage to Broca's area (through stroke or trauma) suffer from a speech disorder, called Broca's aphasia, where they know exactly what they want to say (their thinking is mostly unimpaired), but struggle to articulate it. Trying to say "Pass the salt, please," they might laboriously stammer, "Suh-suh-salt . . . Po-po-pose . . . Pass . . . Puh-puh-please"—often mixing up word order and putting incorrect sounds into the words, like "galt" for salt, or "clease" for please." Psychologist Eric Lenneberg discovered that children between two and ten years old, with injuries to Broca's area, manifest the aphasia. But unlike adult stroke victims, who rarely regain normal speech, the children soon recovered perfect fluency, because their young brains are so "plastic"—that is, they can rewire themselves, myelinating new circuitry, with ease. (So astonishingly plastic is the newborn brain, babies born entirely without a left hemisphere, and thus without a Broca's area at all, briskly rewire the *right* hemisphere for speech and are able to talk normally.)[41]

Children who suffer strokes after age nine or ten have a very different prognosis. "Aphasias that develop from this age on . . . commonly leave some trace behind which the patient cannot overcome," Lenneberg reported, "there will be odd hesitation pauses, searching for words, or the utterance of an inappropriate word or sound sequence."[42] The linguistic brain, he added, "behaves as if it had become set in its ways."[43] Which is why, by the time we enter high school, foreign languages have to be learned through conscious effort ("*je suis, tu es, il est, nous sommes, vous etes* . . ."), and certain accent markers can, no matter how hard we try, be just *aboot* impossible to lose.

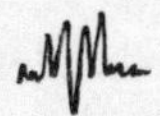

Some of the most convincing (if upsetting) evidence for critical periods in speech are "feral children": those rare cases of babies raised with little or no exposure to human voice sounds. The most notorious modern example is

"Genie," the pseudonym for a grotesquely abused girl held captive, from infancy on, by her deranged father. Strapped to a child's toilet seat, day and night, and isolated in a soundless back room of the family's suburban Los Angeles home, she was not spoken to (except in animal-like barks and snarls) throughout all the critical stages of language development. She escaped her captivity in 1970, at age thirteen but, despite the efforts of speech therapists and psychologists, was able to acquire only a handful of words, which she could not string into sentences. Her voice, through disuse, had also failed to develop properly: respiration, phonation, pitch control, and articulation were all badly stunted. "She had been beaten for vocalizing," Susan Curtiss, a linguist who treated the girl, told author Russ Rymer in 1993. "So when she spoke she was very tense, very breathy and soft. She couldn't be understood. There was a lot of sound distortion, as though she had cerebral palsy, but there was no evidence of muscle or nerve damage."[44] That Curtiss should have likened Genie's distorted speech to that of someone with cerebral palsy (a disease that attacks the basal ganglia) was, in retrospect, prescient: the basal ganglia's critical role in both learning language, and in coordinating breathing, phonation, and articulatory movements, had not yet been discovered.

Today, Genie's permanent mutism remains the most salient, and disturbing, evidence that humans *must* hear human voices talking, within a particular time window, in order to master speech and language.

Because it is unethical to raise a child under twenty-four-hour scrutiny in a language lab, and feral children are (fortunately) rare, most of what we know about the precise stages of speech acquisition derives from baby diaries kept by attentive parents, usually scientists. In the early twentieth century, German psychologist William Stern and his wife, Clara, spent eighteen years obsessively documenting the lives of their three children from birth to around ten years old[45] and provided clear milestones for first cries (at birth), coos and gurgles (at six to eight weeks), babbling (at six to nine months), first words (around one year), and, finally, articulate speech (three years old). But some of the most fascinating insights we possess on language acquisition date to a half century earlier, when Charles Darwin documented the first year in the life of his son, William, born in 1839.[46]

Darwin noted that, by his first birthday, William had yet to utter his first real word—although he did use an invented term for food ("*mum*"). Especially intriguing to Darwin, however, was the prosody, the music, William used when asking for *mum*: "he gave to it . . . a most strongly marked interrogatory sound at the end." (That is, he raised the pitch across the word.) "He also gave to 'Ah,' which he chiefly used at first when recognizing any person or his own image in a mirror, an exclamatory sound, such as we employ when surprised." Darwin, here, recorded a previously unremarked upon fact about how babies learn to talk: the music of speech—its expressive prosody—emerges before words. Tunes before lyrics. This insight would have strong implications for Darwin's later theory of how language evolved in humans (as we'll see); but for now it is enough to note that modern research has confirmed Darwin's observation that a language's specific prosody, its unique melody and rhythm, emerge in a baby's babbling before those sing-songy articulations are molded into specific words. Indeed, an eight-month-old's variegated babbling so closely conforms to the rhythmic stress and pitch patterns of its native tongue that it sounds uncannily like actual speech. You can see this for yourself on YouTube, which is replete with videos of parents jokily "conversing" with their nonsense-spewing infants.[47] But it's no joke: lots of crucial learning is going on during those parent-baby exchanges. And the adults are not necessarily the ones in charge, as was demonstrated in clever experiments of the mid-eighties.[48]

Over large closed-circuit TV screens (this was before Skype, but the participants were, effectively, Skyping), mothers exchanged vocal noises with their eight-week-old babies. Parent and child fell into the expected exchange of musical expressions of attention and affection: the baby's coos, sighs, and gurgles triggering the mother's singsong Motherese, which in turn triggered more coos from the babies, and more Motherese. But when the researchers secretly played a recorded image of the baby made minutes earlier, and the child's sounds were now out of sync with the mother's vocalizations, the women's Motherese vanished. Their voices lowered to normal pitch and their speech took on an adult pace and complexity. The researchers concluded that Motherese, however instinctual, is also part of a feedback loop driven by the *baby's* vocal sounds. In short, two-month-olds conspire with parents in their own language tutoring—and, with their own baby voices, help guide it.

In those exchanges, the baby is also learning an aspect of voice crucial to our species' unique powers of cooperation and goal sharing: *conversation,* a behavior far more complex than you might realize. But think about it. Back-and-forth conversation, although an unplanned activity ungoverned by any prearranged rules of turn duration, topic, or overall length, is amazingly orderly. Indeed, a team of sociologists in the early 1970s[49] showed that when we talk with others, we not only overlap very rarely (only 5 percent of the time), but we speak with almost no gap between when we fall silent and our interlocutor starts speaking. The elapsed time between turns is, on average, 200 milliseconds—too short for our ears to detect, so that conversation sounds continuous, one voice seamlessly picking up where the other left off. But that 200-millisecond gap also happens to be too short for a listener to decode what was said to him and to make a reply. Conversation, in other words, should be impossible.

To understand why I say this, you need to know something about how we, as listeners, turn the incoming ripple of air vibrations that emerge from a speaker's lips into meaningful speech. Because when we hear someone say "Pass the salt, please," we are hearing *not* an "airborne alphabet," not individual auditory "letters," that "spell out" a message. We are not even hearing individual vowels and consonants—or phonemes (the technical term for speech sounds)—which are, after all, only segments of sound that, for the purposes of analysis, speech scientists arbitrarily divide into manageable units (remember when I said that speech is a ribbon of sounds with no gaps in between?). Who is to say where the *l* in "salt" ends and the *t* sound actually begins? In reality, speech is a connected flow of ever-changing, harmonically rich *musical* pitches determined by the rate at which the phonating vocal cords vibrate, the complex overtone spectrum as filtered by the rapidly changing length and shape of throat, mouth, and lips (to produce the vowels), interspersed with those bursts of noise: the hisses, hums, pops, that we call consonants. The effect is not dissimilar—in fact, it's identical—to an instrumental group playing ringing harmonic piano chords and single guitar notes, punctuated by noisy cymbal crashes, snare hits, and maracas: our voices, quite literally, comprise a mini-symphonic ensemble. It is our brain that turns this incoming stream of sonic air disturbances, this strange music, into something deeply meaningful.

Researchers at Haskins Labs in New York City offered the best explanation for how our brain does this when they published, in 1967, their "motor theory of speech perception," which emerged from the conceptual breakthrough that spoken language doesn't begin as *sound*;[50] it begins as a series of dancelike bodily movements, gestures, built from a set of mental instructions originating in Broca's area and another region, a few inches back in the left hemisphere, Wernicke's area, which is where we store our mental dictionary—the words that Broca's area assembles and inserts into sentences. This "mentalese" is only *then* converted into speech by muscle commands delivered to the vocal organs,[51] which produce a complicated soundwave rich in linguistic information. This sound wave is perceived as meaningful speech not by the listener's auditory centers, but by her *motor* centers. That is, the brain circuitry responsible for animating *her* vocal organs is activated by *your* sound wave which triggers, in her brain, an internal, unvoiced, silent, *neural* speech that exactly mirrors the motor instructions you sent to your larynx and articulators, thereby informing her what lip shapes and tongue heights and articulatory targets *you* hit to produce your request. Reverse engineering your vocal sound wave in her brain, she runs it from her motor circuits up through her Broca's and Wernicke's areas to retrieve the vowels and consonants, words and syntax that made up your utterance, and thus arrives at the thought formed in your frontal lobes. And she hands you the salt.

All of which suggests that every act of verbal give-and-take—no matter how hostile, accusatory, contradictory, or aggressive—is also one of extraordinary empathy and intimacy, bringing the brains of speaker and listener into a perfect synchrony, and symmetry, of neural firing. At least, on the level of language. How we communicate, perceive, and interpret the *emotional* channel of the voice, which is superimposed on, or interwoven with, the linguistic channel, is another story altogether, and one that I'll save for later. For now, we must return to my earlier, seemingly nonsensical, claim that conversation should be impossible.

I said so because of the long and laborious cognitive stages your dinner companion had to grind through in order to process the sentence "Pass the salt, please." To make her own verbal reply, she then has to decide on a response ("No—I think you use too much salt" or "Are you sure you wouldn't prefer some pepper?")—a snatch of mentalese she must pass through her language centers and motor cortex to produce the muscle commands that animate her vocal organs to voice her reply. All of this takes, at minimum, about 600 milliseconds:[52] *three times* longer than the 200 milliseconds in which people routinely reply. To respond that fast means that we must be gearing up to speak *well before* the other person has finished talking, and that, furthermore (since we so rarely overlap), we must have a very good idea when our interlocutor is going to fall silent.

Researchers first theorized that we use subtle *language* clues to figure out when a speaker's "turn" will end, as well as reading telltale hand or eye movements that also help us anticipate when our turn is coming. But there was a problem with this explanation: people speak with no-gap/no-overlap precision over the phone, when they can't see each other. And linguistic clues can't be how we orchestrate turn taking because, as we've noted, it takes too long to process what is said to us and then to formulate and execute our reply. Instead, research shows, seamless conversation is possible thanks to the music of speech: the changing pitch of the voice as it follows a sentence's melodic trajectory; the duration of the vowels (which determines the rhythm of an utterance); and the changes in volume from loud to soft. In a 2015 study at the Max Planck Institute for Psycholinguistics and recounted in linguist N. J. Enfield's book *How We Talk: The Inner Workings of Conversation*,[53] researchers asked students at Radboud University, in the Netherlands: "So you're a student?" and "So you're a student at Radboud University?" In the first instance, students responded "Yes" the instant the speaker finished the word "student," with no overlap. In the second, students were equally precise, saying "Yes" after *University*, with no gap and no overlap. They could do this only because they used the music of speech to time their reply. In the shorter question, the pitch on *student* rose sharply, tripling from around 120 cycles per second to 360 cps, signaling a query, and the last vowel was extended (*studeeent*?), giving the listener a head start for dropping in a "Yes" within the 200-millisecond time window. In the second instance, the pitch on *student* didn't rise, and its final syllable was clipped, rhythmic and melodic hints that there were more words to come, cuing the listener to hold off on responding.

The subspecialty of linguistics that studies conversation is called "discourse analysis," and it looks at all aspects of the prosody that guides our verbal give-and-take, including the communicative power of silences. A pause lasting longer than the usual 200 ms between turns suggests that the person about to speak is giving special consideration to her reply, but there's a limit to how long a person can remain silent. Pauses that stretch beyond one second cease to seem polite and create doubt whether a reply is ever coming. Anything longer than a two- or three-second pause is socially unendurable, creating an awkwardness that forces the original speaker to start babbling to fill the silence. Journalists like me take advantage of this impulse by deliberately failing to ask a follow-up question when a subject is clearly struggling to hold back a secret, a tactic as socially excruciating for the reporter as the interviewee.

All of this suggests that conversation is conducted, like a piece of music, according to an agreed-upon "time signature," a kind of internal metronome that determines the number of syllables produced per second. When one speaker deviates from the established rhythm, she signals disagreement. You "hear" this regularly on cable news channels where one speaker's assertion, made in a slow, calm voice is met with a response spoken at high speed—and possibly even interrupting. You don't have to hear what either person is saying to know that they disagree. Vocal pitch operates the same way. Like jazz musicians taking a solo, speakers can improvise freely in terms of content (that is, they can say whatever they want, and at whatever length), but only within a predetermined and mutually agreed upon pitch, or musical "key." A sudden switch in register signals discord.

This tonal dimension of conversation was first described by discourse analyst David Brazil in the mid-1980s.[54] Brazil showed that we converse in three broadly distinct pitches (high, mid, and low). If we disagree with something that someone has said to us, we start speaking at a pitch several semitones *higher* than the voice of the person who just left off. Brazil calls this a "contrastive" pitch. The mid register—in which the person making a reply starts speaking on the *same* vocal pitch as the person who just stopped talking—signals agreement. We reply in the *low* voice, in a register below that of the previous speaker, when agreement is so strong that it is, as Brazil called it, "a foregone conclusion," that is, hardly worth stating.

Conversation, in other words, is not isolated voices making separate points in an alternating series of monologues; it is a creative collaboration, orchestrated not according to linguistic, but *prosodic*, rules. It is a form of singing, a duet in which two brains choreograph, through variations in pitch, pace, and rhythm, the exchange of ideas. Ideally, conversation in our species enhances our advancement through the statement of a thesis, the response of an antithesis, and the eventual arrival at a synthesis—a new idea that emerges from the creative, and civil, exchange of (musically orchestrated) ideas.

Anne Wennerstrom, a discourse analyst, notes that the terms we use to describe successful versus unsuccessful conversations frequently draw on musical metaphors. Good conversations are "in synch," or "in tune," or "harmonious"; the speakers "didn't miss a beat." Bad conversations are "out of synch," "out of tune," "discordant," with the speakers "off their stride" and "on a different wavelength."[55] These metaphors are no coincidence. To a degree that we rarely, if ever, acknowledge, conversation is music—a music learned, and reinforced, in those earliest, seemingly meaningless, exchanges between parents and babbling babies.

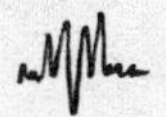

Babies produce their first spoken word around age one, after which they start adding new words at breathtaking speed. In one study, two-year-olds were shown mysterious objects like an apple corer and told just *once* that it is a "dax."[56] Though never again told this word, they recalled it weeks later when asked to point to a picture of the "dax" on a screen. Cognitive scientists call this incredibly speedy word acquisition (and retention) "fast-mapping," and it allows children to stock Wernicke's mental dictionary at the blinding rate of roughly one new word every two hours—a rate they keep up for *fifteen years*, which is how the average child, by the end of high school, has accumulated some sixty thousand words in her mental dictionary. Lest this number fail sufficiently to amaze, consider how difficult it would be to remember sixty thousand telephone numbers or internet passwords. We remember words so well because they are not random assemblages of information units, like phone numbers or passwords. Words carry meaning and we are clearly a species that craves meaning.

The true power of language emerges when babies start combining words, to make complex meanings. They first do this around their second birthday, blurting out two-word utterances, like "All wet!" or "No bed!" or "Play iPad!" Simple as these utterances seem, they betray a remarkable syntactic sophistication, including an understanding that word order affects meaning ("Lucy hit" means something different than "Hit Lucy"). Chomsky argues that word order could not be learned from the "impoverished" input of half-heard, rapid-fire, degraded adult speech, and so must be inborn. But linguists like Snow, Ferguson, Garnica, and Fernald have shown how slow, repetitious, singsong Motherese teaches exactly the kind of telegraphic two-word utterances, heavy on word order, that toddlers end up speaking as their first grammatical constructions.

Kids seem to stall out in the two-word stage for a long time—often up to a year. They might add a word or two to the speech stream ("Lucy hit doll" or "Go out play!"), but their speech remains telegraphic, choppy. Then, around their third birthday, they magically start speaking in fluent multi-word sentences. Again, Chomsky saw this as proof that language must be inborn, and Steven Pinker, in *The Language Instinct* (a popularizing explanation of Chomsky's theory), reinforced this idea in his description of a boy named Adam who, at age two, is limited to utterances like "Play checkers" and "Big drum," but who, suddenly, at age three, starts saying things like, "So it can't be cleaned?" and "Can I put my head in the mailbox so the mailman can know where I are and put me in the mailbox?"[57] Pinker calls Adam's "explosive" linguistic near-mastery evidence that grammar must be preinstalled before birth.

But as we've noted, Adam has, over the previous *three* years, been enrolled in as intensive a language immersion course as can be imagined: all day, every day, with no time off on weekends and no breaks except for naps. You could earn an undergraduate degree in astrophysics in that time—and you, as an adult, don't learn with anything like the speed of an infant. That Adam's linguistic aptitude *appears* sudden is because the bulk of his learning took place before he could do much more than coo or babble.[58]

For Chomsky and his acolytes, the definitive evidence that our brains come preinstalled with syntax and grammar is the sheer *complexity* of three-year-old Adam's speech—a complexity, Pinker says, that "could not have been taught."[59] He furnishes a famous example from Chomsky, who analyzed how small children form a question from the statement, "A unicorn is in the garden." Even a three-year-old will know that to make a question you shift the verb ("is") to the front ("*Is* a unicorn in the garden?"). But Chomsky warned against assuming that children do this thanks to a rule learned from hearing adults speak—a rule such as: *To form questions, scan the sentence for the verb and slide it to the front.* Because if you apply that rule to a more complex sentence like "A unicorn that is eating a flower is in the garden," you produce nonsense: "Is a unicorn that eating a flower is in the garden?"

Chomsky correctly pointed out that children never make this error. With the more complex sentence, they somehow know to skip over the first "is" and shift the *second* "is" to the front—proof, in Chomsky's eyes, that very small children innately understand the "deep structures" of Universal Grammar—language not as a string of words, but as a set of interlocking chunks made up of verb phrases like "is eating a flower," and noun phrases like "a unicorn that is in the garden." Matt Ridley, an award-winning science writer, also used Chomsky's "unicorn" example in his bestseller *Genome* to argue (like Pinker) that syntax must be inborn because moving the second "is" to the front of the sentence indicates a knowledge of noun and verb phrases that "could not be inferred from the examples of everyday speech without great difficulty."[60]

I beg to differ. It is precisely in "everyday speech" that children hear and learn the distinct noun and verb phrases in sentences like "A unicorn that is eating a flower is in the garden." Those phrases are distinguished by dramatic changes in pitch and rhythm across the utterance, by the melodic changes we use to help people follow what we're saying. Not only do we lower our pitch after "unicorn," to sonically tuck one phrase ("that is in the garden") into the other, we also slightly increase the speed of articulation for the embedded chunk, so we don't put undue demands on the listener's short-term memory; we want her to remember that we're talking about the unicorn mentioned in the earlier part of the sentence. If you doubt the rich melodic and rhythmic elasticity of the utterance, try saying it on a single note and at a steady unchanging pace. "A. Unicorn. That. Is. Eating. A. Flower. . . ." You will sound ridiculous, as if trying to imitate a robot.

For three-year-olds who have been making exceptionally subtle discriminations in the voice's fundamental frequency *since the womb* (and for several years after birth), detecting the significant drop in pitch across the embedded phrase is achieved not, as Ridley states, with "great difficulty"; hearing that change is (pun intended) child's play. Which is why toddlers would never say the discordant, melodically impossible, and rhythmically ungainly utterance: "Is a unicorn that in the garden is eating a flower?" It *sounds wrong*. Why? Because the human voice speaking—*the music of speech*—teaches infants how sentences are put together, where sentences properly begin and end, how noun and verb phrases are embedded, and how they can be shifted around to maintain the satisfying, songlike, melodic, and rhythmic resolutions that characterize all human speech.

In the book *This Is Your Brain on Music*, musician and neurophysiologist Daniel Levitin describes what happens when melodies resolve; that is, when a tune satisfyingly arrives back at the note that established the melody's key. This actually causes the release of dopamine in the brain, giving the listener a pleasurable sensation of reward.[61] Babies have been gleaning those pleasurable rewards, from spoken language, since they were in the womb, and they use those rewards to learn the underlying grammar (Chomsky calls them *syntactic structures*) of the language they hear spoken all around them.

This was demonstrated in elegant experiments using four-month-old babies. Temple University linguist Kathy Hirsh-Pasek played speech samples through speakers placed on either side of a baby. Infants indicate their interest and attraction to particular sounds by turning toward the speaker broadcasting the sample. Hirsh-Pasek first played an audiotape of a woman reading a storybook passage:

> Once upon a time, a lady and a witch lived in a big house. The house was very old and messy. It had a big garden and six windows in the front.

As four-month-olds, the babies couldn't understand a word, but hearing speech that conformed to the prosodic patterns they had become accustomed to since the womb, they turned to face the speaker broadcasting the story, clearly delighting in the dopamine rush of all those nicely resolving vocal beats and melodies. But Hirsh-Pasek also created a doctored version of the same tape, in which she inserted one-second pauses in random places:

> Once upon a time, a lady [*one-second pause*] and a witch lived in a big house. The house was [*one-second pause*] very old and messy. It had a big garden and [*one-second pause*] six windows in the front.

The babies showed their confusion and distaste by turning away from the speaker.[62] Interrupting a melody mid-arc, and then resuming at the unnatural pitch of a voice in midsentence, created discordances of timing and tune that the babies could not tolerate—strong proof that long before they could ever dream of assembling a complex sentence filled with properly interlocking noun and verb phrases, a baby has learned, through close attention to the voice, how those "syntactic structures" work: because of how language *sounds*.

Speech scientists call these grammar-defining vocal melodies *linguistic* prosody (in contrast to *emotional* prosody) and it is controlled, processed (and perceived) by brain areas quite distinct from the left-brain "language areas"—indeed, in the opposite hemisphere, in those parts of the *right* side of the brain associated with the making of, and appreciation for, music: pitch, rhythm, pace. We know this because stroke patients who experience damage to the right side of the brain often speak in a monotone—and are incapable of hearing the linguistic prosody in others' speech. The drastic limits that this places on such patients' ability to understand, and be understood, leaves no doubt of the crucial role played by the music to which we set the lyrics of everything we say. (Even, as we saw earlier, as mundane a two-syllable utterance as "Hello.")

While recent science has provided startling insights into how we, as individuals, acquire language in infancy, debate rages over how that skill was acquired by our species as a whole. To believers in the Great Leap Forward, our speech is a behavior *discontinuous* with our animal lineage and thus represents the most abrupt, and dramatic, of behavioral breaks from our nonhuman past. To others, like Darwin, our speech is *continuous* and grew out of the vocal noises of our evolutionary ancestors, as a refinement of the emotional vocalizations that every mammal and bird uses to drive off enemies and woo mates—a view with the widest possible implications for our species, since it establishes the voice as *the* most conspicuous borderline between our nonhuman and human identities, as the single biological endowment where both aspects of our nature are fused in a single act: an acoustic signal that at once embodies our most extraordinary and defining attribute (speech), yet delivered by a mechanism we share with every bird and mammal, no matter how "bestial" or primitive. Which has had (to put it mildly) big repercussions not only for how our voice affects our private, personal relations with friends and family, but for how the voices of public figures guide the fates of societies and civilizations.

My allegiance is to the latter, Darwinian, view. To understand why, we must go back in time. Back to when the voice began.

TWO

ORIGINS

The voice of every animal (including birds, dogs, lions, sheep, seals, frogs, cats, chimpanzees, mice, us) shares at least two traits in common: they are sounds powered by the lungs and emitted through the mouth; and every voice (barks, whinnies, whines, chirps, squeals, meows, ribbits, roars, the State of the Union address) derives from a common ancestor, an animal we don't ordinarily associate with voice: fish.

To understand how this could possibly be so, we must travel to a time around 530 million years ago, when the first fish evolved. Like their living descendants, these ancient fish sustained life by extracting oxygen from the water and expelling CO_2 with a specialized membrane that lines the inside of the throat: gills. Some of these primordial fish, however, evolved in shallow lakes or swamps and during droughts would become stranded on land. Many suffocated to death, but at least one was lucky enough to undergo one of those random mutations that drive natural selection.

In this case, a possible copying error in one of the genes responsible for building gills, rendering the subtly altered membrane capable of pulling a little oxygen from the *air*—a tiny sip that kept the landlocked fish alive long enough, not only to survive the dry spell, but to mate and pass along the mutated gill gene and the tiny survival advantage it conferred to its offspring. Over hundreds of thousands of years, and many other random mutations that improved the animal's ability to survive on land, a new species evolved in these swampy, shallow-water areas, a transitional, hybrid animal that possessed *both* water-breathing gills *and* rudimentary air-breathing lungs, which had formed from the hollow swim bladders it used for flotation. These creatures are known as lungfish, and they are our oldest air-breathing, land-dwelling relatives.

They can still be found in the swamps of South America, Africa, and Australia: remarkable animals so little changed from their ancient ancestors that they are known as "living fossils." Darwin, in the *Origin of Species*, used the lungfish to illustrate a central concept of evolution by natural selection: namely, that "an organ originally constructed for one purpose" (the swim bladder, for flotation) "may be converted into one for a wholly different purpose" (the lung, for respiration).[1] Thus did Darwin document a crucial milestone in the origin of what would become our human voice: the emergence of the air-propelling bellows that powers our speech and song. It was, however, up to another scientist, writing seven decades later, to reveal how another key adaption in the lungfish gave rise to the voice.

Victor Negus was a thirty-four-year-old First World War veteran, in 1921, when he began a residency in throat surgery at King's College Hospital in London. There, he undertook a research project on "the production of voice in animals and man."[2] The planned two-year thesis stretched to nine years, as Negus dissected an ever-growing menagerie that included fish, lizards, frogs, birds, and various mammals. From seeking to learn *how* the voice is produced, Negus found himself on a quest for *where* it had come from. The resulting five-hundred-page treatise, *The Mechanism of the Larynx* (1929), became the most important reference work on the subject for the next half century and the basis for his eventual knighthood. As Negus showed, our voice starts with the lungfish.

He introduces the animal on page three, where he describes the dissections he performed on an Australian species called the *Lepidosiren.* He notes how a hole had developed in the gills that separated the throat from the digestive tract, creating an opening in the bottom of the mouth that led into the swim bladder, whose lining had thinned to the point where oxygen could diffuse through the membrane into the blood vessels beneath: the primitive air-breathing lung. No more suffocating when landlocked. But the hole in the animal's throat also left the creature vulnerable to drowning when it returned to its aquatic existence. "Therefore," Negus wrote, "it became imperative that only air, and not water or other harmful substances, should enter [the lung]. With this object in view, a valve was evolved to guard the entrance to the pulmonary outgrowth."[3]

As the image of *my* throat projected on Dr. Woo's computer screen makes clear, our vocal cords are an inheritance from these ancient fish—a *valve* that opens and closes over the opening to our windpipe and that we hold in the open position to allow air to pass to and from our lungs (as we breathe), but which we snap closed over the windpipe when "water or other harmful substances" threaten to enter our lungs and choke us to death—or when we wish to make voice sounds. Air pushed up from the lungs encounters the barrier of the closed vocal valve which makes the membranes flutter and flap against each other in the same way that your lightly sealed lips flap noisily against each other when you blow a Bronx cheer.

The term vocal "*cords*" is thus a misnomer. It dates from the mid-eighteenth century when the French anatomist Antoine Ferrein, studying the larynges of animal and human cadavers, compared the membranes to "violin strings" (*cordes*) that vibrate under the "bowing action" of the breath.[4] Our vocal cords don't produce sound that way. As a valve set in motion by air from the lungs, the vocal cords actually *chop* the airstream into rapid pulses that produce a sound wave. This distinction helps explain the vocal injury that Adele, Julie Andrews, and I sustained (singing a high, loud note, we can clash the vocal cords together up to a thousand times a second). It also has relevance for the fate of our species—since the raw sound source of the air-chopping vocal cords creates a particularly rich and buzzy sound wave with lots of overtones—those higher frequencies that we filter with movements of the tongue and lips to make the vowel formants that distinguish *head, hid, hood, had,* and so on, and that endow us with the ability to produce clear and intelligible speech.

For the lungfish, vocal sounds are about as far from articulated speech as can be imagined: squeezing air through the simple sphincter of its throat valve, the animal can produce only an array of grunts, squeaks, hisses, and belches (I am trying to avoid the word "farts" but I'm afraid those were the first vocal sounds heard on Earth). Over millennia, however, the larynx underwent drastic physical changes as the lungfish's land-dwelling descendants (first amphibians, then reptiles, and finally mammals) refined the vocal valve's efficiency for breathing—and for the production of voice sounds so useful for survival and reproduction. These changes included the addition of a system of movable cartilages to which the ends of the vocal membranes attach, and a complex musculature that permits a stretching or slackening of the vocal cords to vary vocal pitch (pull them taut and they chop the air faster, raising the pitch; loosen them and they chop the air slower, giving a lower note), and to stiffen the membranes to create sounds like a growl.

Changes in the voice's volume, from loud to soft, are controlled by the speed and force with which we drive the air up through our vocal cords—a refinement of the respiratory system that appeared in mammals, around 220 million years ago, with the emergence of the diaphragm, a thick muscle that attaches to the bottom of the rib cage and that divides the trunk into an upper chamber, housing the lungs and heart, and a lower chamber, housing the stomach and intestines. Movements of the diaphragm up and down control the action of inhaling and exhaling, and fine motor control of the speed and force with which the diaphragm propels air from the lungs is how we and other mammals vary the loudness and softness of vocalizations. But it was yet another evolutionary enhancement unique to mammals that would have the most decisive impact on our human voice. Mammals are defined as a species by their habit of feeding their babies with milk from the mother's mammary glands (the term "mammal" is from the Latin *mammalis*, "of the breast"). Our baby mammal ancestors, by affixing their lips to a teat and performing a suite of complex sucking and swallowing maneuvers, developed the throat, mouth, tongue, and facial muscles that our species would learn to coordinate for articulated speech.

Most mammals, then, possess all the vocal apparatus necessary for talking. Indeed, a chimp's lips, tongue, velum, lungs, and larynx are virtually indistinguishable in structure and function from ours. Likewise, the rest of its anatomy, from its frontally placed eyes, opposable thumbs, two symmetrical nipples, and shortened snout. So anatomically similar to us are apes that the eighteenth-century Swedish naturalist Carl Linnaeus classified humans and all simian species in the same mammalian order, which he called "primate" (from the Latin *primus* for "first rank"). Working a century before Darwin, Linnaeus drew no *evolutionary* connection between apes and us. He was going purely on anatomical similarities. It was only because of concerns expressed by the church (which said that man was created in the image of God, and apes, by logical inference, weren't) that Linnaeus eventually invented a separate primate classification that lifted us above the rest of the animal kingdom in a genus he called *Homo* (Latin for "Man") and a species he called *sapiens* ("wise"). However, Linnaeus continued (privately) to insist, in letters to biologist friends: "I know scarcely one feature by which man can be distinguished from apes."[5] Well, there was *one* clear distinction—although it was, he said, a *behavioral* rather than an *anatomical* difference.

The human capacity for speech.

Linnaeus was a few centuries ahead of his time in recognizing that our unique ability for spoken language resides not in any dramatic anatomical difference between us and our nearest animal kin—except for our slightly lower larynx, a unique distinction of our species that Linnaeus never mentioned, no doubt because he considered it unimportant, but whose implications for the origins of language we will look at later. For now, it is enough to point out that our capacity for speech derives, chiefly, from differences in our brain, especially the cortex, the wrinkled outer layer where Broca's and Wernicke's areas are found.

However, it is also vitally important to stress that, when we speak, we don't draw solely on the cortex and the activity in Broca's and Wernicke's areas. If we did, we would sound like HAL in *2001*, or Mr. Spock in *Star Trek*: precise, but affectless, flat, weirdly inhuman. ("A. Unicorn. That. Is. Eating. A. Flower. . . .") What makes our voice human (what makes *us* human) is emotion—the feelings and moods that underwrite everything we do and that emerge, vocally, in our prosody. Nor is this music a mere garnish, or add-on; we've already seen how the melodic and rhythmic contours of speech orchestrate conversation, define syntactic and grammatical units, and, indeed, inculcate language in newborns. It was from this music, Darwin says, that our capacity for speech evolved in the first place, through a shaping of emotional, expressive vocal sounds. To appreciate how this came about, we need to glance back over our evolutionary history, to track developments that took place in parallel to the bodily changes that occurred as the various speech organs (lungs, larynx, lips, and tongue) were evolving—to events unfolding inside the skull.

THREE

EMOTION

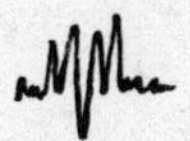

Over the course of animal evolution, three more or less distinct layers of the brain emerged,[1] layering themselves one on top of the other. We retain all three layers, each of which plays a specific role in how we control (or fail to control) the emotional channel of our voice.

The oldest layer, the brainstem, which first began to develop with the marine ragworm over 600 million years ago and was refined in the earliest fish, is the seat of all our involuntary processes like breathing, blinking, and heartbeat. Reptiles possess a brain that is almost all brainstem and their existence is duly characterized by largely instinctive, reflexive behaviors devoid of anything we would describe as "emotional." Lizards don't even form maternal attachments to offspring—they will eat their babies after they emerge from the egg (newly hatched lizards instinctually flee to treetops to avoid being cannibalized). Because the voice, in all species, is first and foremost an instrument of communication, of *social* interaction, the lack of sociality in reptiles had huge implications for how their voices evolved—or, more to the point, didn't. Of the six thousand or so lizard species, most don't vocalize at all, even to repel aggressors or express an urge to mate (instead, they use silent body postures and tail wags). You *can* get a lizard to use its voice, by accidentally stepping on it, which calls up a pure pain response in its brainstem. The rest is silence.

Our brainstem, which we inherited more or less intact from reptiles, behaves exactly like a lizard's, vocally speaking. You can test this next time you accidentally hit your thumb with a hammer. You will make a noise that issues directly from your brainstem or, as it is colloquially known, your "lizard brain"—a scream that might sound like *Oww!* or *Augghh!* or perhaps *AHHHH!*—it's impossible to render in writing for the very reason that such noises are made without access to the vowel-and-consonant-generating activity of Broca's and Wernicke's areas up in the newfangled cortex. Speech scientists call these noises "fixed action patterns." They're hardwired into the brainstem and activated by an "innate releasing mechanism"—in the hammer-to-thumb example, the pain response. They can be triggered by other kinds of extreme sensory input, like the pleasurable sensations that lead to orgasm (moans or excited cries), or the shout you produce upon entering what you assume to be your empty home only to have the lights flash on and people scream "Happy Birthday!"—a surprise sensory assault that stimulates the visual and auditory inputs to your brainstem with sufficient alarm to provoke a reflexive, fixed-action *Urrgh!* or *Aaaukkkh!*

For very young humans, the surprise sensory assault is the sudden bright light and ear-splitting noises of the delivery room—which, as noted earlier, make them reflexively go *Waaaah!* That this noise is a hardwired, preprogrammed reflex is clear, not only from a newborn's ability to produce it upon first contact with air, but from a tragic "experiment of nature"—a rare birth disorder in which human babies are born with nothing *but* a brainstem, their higher brain layers having failed to develop. With so severe a deficit, it might be supposed that these children, who live only a few hours or days after birth, would be vocally silent. Instead, they react to pain with perfectly normal-sounding baby cries.[2]

Though involuntary sobs, laughter, and cries of pain or pleasure all express emotions, voice researchers don't classify them as emotional signals. They call them *interjections*, and afford them about as much interest as the kick of your foot when the doctor taps your knee with his rubber hammer.

Of far greater fascination to voice science are the more subtle vocal emotions, the prosodic shadings that betray anxiety, hostility, lust, doubt, guilt, love. These sounds are generated in the second evolutionary layer of the brain, which emerged with mammals. Known as the limbic system, this layer grows over top of the brainstem and it generates emotions in us by triggering the release of chemicals called neurotransmitters that flood our body, producing inner states of feeling that aid us in reacting appropriately to the situations we encounter. Thus the sight of a menacing-looking stranger advancing upon you in a dark alley activates a limbic structure called the amygdala, a small kidney-shaped organ (we actually have two of them, perched on either side of the brainstem), which signals to the endocrine glands for the release of adrenaline, which kicks up your heart rate, speeds your breathing, induces sweating, and prepares your muscles to fight or flee—and sends a feedback signal along another neural pathway to the higher brain centers that sparks the conscious, "felt" emotion of terror. By contrast, the sight of your newborn baby smiling up at you acts on a different limbic structure (the nucleus accumbens), an anticipatory-pleasure and reward center, which triggers release of the neurotransmitters dopamine and serotonin, endogenous opiates that your own body's glands produce to suppress physical pain. These chemicals create subjective "felt" states like contentment, serenity, love. Which is why lizard moms, lacking a limbic system, instead see their babies solely as a way to appease their appetite.

Crucially, the limbic system, as the part of the brain that makes us social animals, also shapes how we *signal to others* our inner states, as a way to enhance our chances of surviving and of mating. These social signals include facial movements like smiling when we encounter an old friend, or furrowing our brow and setting our jaw when we confront a dangerous enemy, or the myriad shadings of emotional prosody in our voice.

The first hard evidence of the limbic system's role in controlling vocal emotion arose from research by the Swiss neuroscientist Walter Rudolf Hess, winner of the Nobel Prize in Physiology or Medicine in 1949. Hess used cats as experimental subjects, surgically implanting hair-thin electrodes into the brain.[3] When they awoke from the operation, the cats could move freely around Hess's lab. By delivering pulses of low-voltage current, he induced "natural" neuron activity in specific areas of the cats' limbic systems and found that, like a gamer using a joystick, he could elicit a wide repertoire of emotional vocalizations in the animals. Activate the amygdala and they produced a hissing, spitting, and snorting, which accompanied classic feline "rage gestures" (dilated pupils, flattened ears, raised fur, extended claws); stimulate the nucleus accumbens and the cats made a "happy purring," which was paired with outward signs of calm and contentment.

In the late 1960s, a German research team at the Max Planck Institute of Psychiatry in Munich adapted Hess's method to study the voice of a higher mammal species, the squirrel monkey, a small, intelligent, exceptionally social primate from the rain forests of South America, who produce a far wider array of vocal noises than do cats: threat calls, mating cries, friendly signals, parental sounds. The team elicited some fifty clearly differentiated emotional calls and cries by stimulating many areas of the monkeys' brains, including specific limbic structures that are also found in our brain. "Growling" indicated a "directed aggression"; "cackling" signaled high excitement; "chirping" was a friendly sound promoting "group cohesion"; "trilling" was nonaggressive and focused group attention; "quacking" expressed "irritation and unease"; "shrieking" signaled the "highest degree of excitement." To no one's surprise, stimulating the amygdala produced the most aggressive sounds, a loud hissing and growling—sounds identical in acoustic profile to those made by cats in experiments in which *their* amygdala was stimulated—"a remarkable homology," the German team noted.[4]

"Homology" refers to the shared ancestry of anatomical structures. The homology was remarkable because the identical, amygdala-driven threat sound, in cats and monkeys, was strong evidence for the evolutionary roots of particular emotional vocalizations in specific limbic structures, structures conserved through all mammalian species, including our own. Which was big news. Because if primates (monkeys) inherited their emotive calls from lower mammals (cats), it was highly likely that we inherited those same calls from our primate forebears.

Charles Darwin was the first to raise this possibility, a century earlier, in his third and final work on evolution: *The Expression of the Emotions in Man and Animals* (1872) where he shocked the world by stating that our emotional expressions do not derive from the stirrings of a divinely bestowed soul, but are instead evolved traits that create all the facial expressions, bodily postures, and vocal sounds "which we recognize as expressive." In this, Darwin said, the voice is "efficient in the highest degree," acting, in hostile encounters, like a "harsh and powerful" weapon to repel antagonists, and in mating and child-rearing, as a soft, high-pitched, loving signal to attract mates or nurture offspring.[5] From these insights, Darwin derived his "principle of antithesis," in which opposite emotions give rise to diametrically opposed body postures, facial expressions, and vocal noises—all part of a social signaling system for repelling predators or winning mates. Aggressive or angry animals stand tall and stiff, raising their hair to suggest a bigger, more formidable body—visual signals that they match with the voice, through low-pitched, growling sounds that also "suggest" a bigger, more formidable body (a size bluff); submissive or loving animals win romantic partners by loosening their stance, even pushing the body against the ground like an affectionate dog, flattening the hair—and producing the high, or whining, pure vocal notes that suggest a small, submissive body (a reverse size bluff).

Darwin's theoretical musings were given empirical support in the 1970s by zoologist Eugene Morton, who used a device called the spectrograph to minutely parse the voices of some fifty-five bird and mammal species, showing that all animals vocalize with low-pitched growls for aggression and high-pitched pure notes for friendliness and mating.[6] (Our current parakeet, Rudy, does this, growling a low, stuttering warning when I extend my finger for her to hop onto, but unfurling mellifluous pure-toned song-twitter when happily perched on my shoulder.) Morton, moreover, showed that animals produce these vocalizations on a graded spectrum between the two extremes of low-pitched growl and high-pitched pure tone, often mixing elements of the two opposed signals in a single bark, chirp, or bleat when unsure of exactly *how* they feel about a potential friend or foe. They trade these vocalizations back and forth, testingly (you've probably noticed dogs doing this with other dogs, moving from submissive, questioning whines to assertive barks to a warning growl—and if all such diplomatic interchanges fail, outright fighting)—a form of animal "conversation" that, if less exquisitely timed than our own, is nevertheless conducted according to similarly coded variations in pitch, volume, and rhythm; by prosodic cues.

Indeed, this basic signaling pattern, essential to the main engines of natural selection (survival and reproduction) was, scientists now believe, hardwired into the animal genome over millions of years, and it sets the basic parameters of emotional prosody. According to linguist John Ohala, this pattern is heard even in the rise in pitch we use at the end of a question, a prosodic feature found in virtually all languages. Ohala says a speaker raises his pitch to signal that he is deferring to his interlocutor, assuming a posture of submission to their greater authority. When speakers answer a question, they do so in a downward pitch trajectory, affirming, vocally, who's top dog.[7]

A critical point in Morton's study of animal voices was that the number and variety of emotional vocalizations, along the spectrum between the endpoints of growl and whine, increases with a species' "intricacy of social behavior." Those animals who live in larger cooperative groups, with more complex social interactions, produce and process signals with more points—more sounds—along the emotional spectrum. Morton cited the German study of the highly social squirrel monkey and its over fifty finely differentiated calls and cries.

We are a species of ape who displays the most varied vocal emotional signaling of any animal, by far. Between the end points of low-pitched, growling anger, and high-pitched, singsong notes of friendly greeting, we produce a near-endless array of vocal inflections and intonations: subtle manipulations of pitch, rhythm, timbre, and volume that can express basic happiness, sadness, rage, or fear, but also such (presumably) uniquely human states as pride, wistfulness, nostalgia; or indeed, nostalgia-tinged-with-aggression, or *Schadenfreude*-with-a-slight-admixture-of-guilt. This subtlety reflects not only the complexity of the social interactions we have to negotiate, but also the labyrinthine variety of our *inner* states of emotion and consciousness, which arise not only in response to events in the present (as is the case with nonhuman animals whose vocalizations are tied to in-the-now reactions to threats, predators, mates, food gathering, and territorial defense), but our *memories* of past acts and encounters, as well as our fears and hopes for the future, and any number of other conscious (and subconscious) thoughts that, as far as science has been able to divine, is a result of our hugely expanded cortex—the third layer of the brain that grows atop the brainstem and limbic system.

When it comes to cortex, size matters. More cortex means more computational power, greater intelligence, and stronger powers of reasoning. Our cortex, which went through a truly explosive period of growth after we parted evolutionary company with chimpanzees some seven million years ago, is by far the biggest in all of nature, three times that of the next largest (the chimp's).

Scientists have offered many explanations for why our cortex expanded so much and so fast. Eugene Morton was the first to propose that emotional vocal signaling was one of them, since perceiving and processing fine differences in vocal pitch and timbre—and assigning to them nuanced gradations of aggression or submission—are acts that require higher-order processing than the brainstem or limbic system can offer. This seems as good a theory as any, and it helps explain why the highly social and vocal squirrel monkey has a cortex three times that of a cat's and why we have a cortex that is, in relation to body size, the biggest of any animal by a long shot.

One way that our massive cortex affects our emotional vocalizations is by editing, or censoring, them—modulating the spontaneous noises that might otherwise burst from us. So, in a hostile encounter with a boss, teacher, parent, spouse, child, or the guy who cuts in front of us in line, we might experience a flare of activity in our amygdala, which ordinarily would trigger a loud, angry growl or hostile snarl. But to blare forth with such a noise would usually come at too high a social cost, so we control the vocal noises we make, keeping the overt hostility out of our tone, and preserving relations with our family, friends, and the line cutter.

Such top-down control of our emotional behavior is an actual physical process in the brain, as first demonstrated in the 1990s by Antonio Damasio, a neuroscientist at the University of Southern California. He used fMRI brain imaging to show that, when emotionally aroused, we send nerve signals along the axons that extend from our cortex down into the limbic structures beneath to subdue limbic activity, dampening or modulating unfiltered emotional responses.[8] The notion that higher cognitive functions (thought, reason, and will) rule over our "animal passions" has informed Western thought for millennia, from Shakespeare's plays to Freud's psychoanalysis (where the civilizing "superego" strives to control the unbridled, animal "id").[9] Damasio showed that this model of human consciousness is not a metaphor: it is a fact—and one reflected in the prosodic contour and color of every syllable we speak.

That our cortex and limbic system operate somewhat independently of each other in our emotional voicing is clear from disorders like Tourette's syndrome in which some sufferers make uncontrolled outbursts that sound like barks and whinnies, neighs and roars. Brain scans reveal overactive amygdalae—and a strangely quieted cortex.[10] That some Tourette's sufferers involuntarily shout swear words—"Shit!" "Fuck!"—would seem to make little sense because language is not computed in the limbic system. But brain scans also show that such "prohibited" words originate, not in the language-generating cortex, but in the limbic emotion centers. "Bad words" have, in effect, been banished from Wernicke's mental dictionary through early social conditioning ("Now, Susie, that's not a nice thing to say!"), and they take up residence in a lower part of the brain, on the margin where the cortex meets the limbic structures.[11] This is why, when you hit your thumb with a hammer, you sometimes let fly with a "*FUCK!!*" or "*SHIT!*" You are accessing these socially prohibited words from the same part of the emotional brain from which people with Tourette's syndrome pluck them.

The independent roles that the cortex and limbic system play in voice has also been beautifully illustrated by neurosurgeons who, when plotting a route through delicate brain tissue to reach a buried lesion, often use an electrode to stimulate tissues under the cortex to determine their function (so that they know what *not* to cut or mangle for fear of causing blindness, deafness, paralysis, or death). The patients remain wide awake throughout (since the brain has no pain receptors) and are able to answer the doctor's questions about bodily sensation or movement elicited by the electrode. When doctors get deep into the brain and stimulate the limbic system, particularly the amygdala, patients behave exactly like Hess's cats, producing cries of raw fear or anger, laced with what one neurosurgeon called "uncontrolled swearing."[12] The instant the electrode is withdrawn, their utterances return to polite G-rated calm.

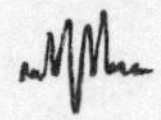

Vocal emotion is not shaped purely by a top-down policing action of the cortex, though. The "animal" limbic system has nerve axons of its own that project upward into the cortex and thus (in the parlance of neuroscience) "talk to" the higher brain regions. Damasio mapped these circuits and showed that emotions, gut feelings, instincts, and intuitions play a far greater role than previously realized in "rational" decision making. To be fully human, we *need* emotions, which, after all, evolved to help us stay alive and pass along our genes. But sometimes our emotions override the rational brain entirely. When this kind of mutiny occurs, the results, in terms of voice, can be calamitous, giving rise to an array of extraordinary, emotion-based vocal disorders.

One of the more dramatic is dysphonia, a clenching of the vocal cords across the windpipe that restricts phonation, producing a strangled sound.[13] Acute sufferers lose the power of speech altogether. Brain scans show that dysphonic patients have highly aroused amygdalae and (as with Tourette's patients) quieted cortical areas that ordinarily damp down amygdala response. In some cases, a single, emotionally cataclysmic event sets the syndrome off—as with the nearly four-decades-long episode that interrupted the career of British singer Shirley Collins, a leading figure in the folk music revival of the 1960s. On albums like *The Power of the True Love Knot* (1968), Collins conjured a delicate, unschooled soprano that she would lift into a spine-shivering falsetto. But in the late 1970s, when she was approaching forty, Collins appeared in a musical in London, performing alongside her husband, Ashley Hutchings, bassist for the folk band Fairport Convention. Early in the play's run, Hutchings announced that he was in love with another actress in the play, and was leaving Collins. Sharing a stage, night after night, with her sexual rival, Collins suffered a trauma that centered, perhaps unsurprisingly, in a part of her that had always been preternaturally expressive of her emotions: her voice. "My voice just—my throat locked," she later said. "Some nights I could manage a few notes, sometimes nothing came out at all when I opened my mouth." Diagnosed with dysphonia, Collins abandoned singing and remained silent, as an artist, for the next two decades, taking odd jobs to support herself and her two children.[14]

Not all emotional vocal disorders result in shutting down the voice. Puberphonia is a syndrome that befalls men in their late teens and early twenties, after they have undergone all the normal bodily changes associated with puberty—including the enlargement and thickening of the vocal cords that ordinarily deepen the voice. Puberphonic voices, however, don't deepen; they retain their high, light, childlike pitch and timbre. In rare cases, a hormonal problem is the cause (the vocal cords, unresponsive to testosterone, fail to enlarge). But the malady is more often psychological—which is to say, emotional—in origin. The puberphonic patient unconsciously pulls the larynx high in the neck, tilting it forward, which both increases tension on the vocal cords and reduces the size of the throat's resonating chamber, giving the voice a high, childlike pitch and timbre. In therapy, sufferers report a fear of growing up, leaving childhood and assuming adult responsibilities. Michael Jackson was believed to be a lifelong sufferer.[15]

Such "bottom-up," limbic influence on the voice is not confined to dramatic disorders like Tourette's, dysphonia, and puberphonia. Activity in the limbic system leaks into the most mundane speech, all the time, for better and for worse. Someone asking for the salt from a dinner companion will manifest entirely different emotional voicings, depending on whether the speaker is talking to a spouse of thirty years, a work rival, or an alluring first date. Indeed, keeping emotion out of the voice can be really hard. In the fall of 2008, during the global subprime mortgage meltdown, I spent a whole day in my home office, unable to work as I absorbed news of the looming apocalypse. Around 3:30, I heard my nine-year-old son arrive home from school (we couldn't see each other; my office is down the hall and around a corner from the front door). Eager to shield him from my anxiety, I called out in a voice that I tried to make sound like my normal, cheerful, after-school greeting: "Hi! How was school?"

His response: "*What's wrong?*"

As this would suggest, humans have evolved exquisitely well-tuned hearing for detecting when someone is attempting a vocal masquerade. Evolutionary biologist Richard Dawkins and zoologist John Krebs, in a now classic 1978 paper, suggest how. They point out that deceptive signaling is, itself, an evolutionary adaptation, a trait that developed in our earliest animal ancestors, to reap survival and reproductive benefits.[16] (We've already seen how hostile mammalian and avian vocalizations are built upon size bluffing through lowered pitch and noisy growling—a "dishonest signal.") Dawkins and Krebs said that such false signaling is found in *all* animal communication: the colors flashed by butterflies, the calls of crickets, the pheromones released by moths and ants, the body postures of lizards, and our acoustic signals. Nature is deceitful. Creatures will do what they can to *not die*—at least until they've succeeded in winning a mate and passing along their genes.

But at the same time, Dawkins and Krebs tell us, the *receivers* of deceptive signals—the would-be dupes—undergo their own coevolutionary "selection pressure" for *detecting* false communications. The coevolution of voice and ear (which began after the lungfish emerged onto land and began to adapt its underwater hearing apparatus for the detection of airborne sounds) initiated a biological "arms race" (in Dawkins and Krebs's memorable phrase). The "manipulating" vocalizer evolves, over vast spans of evolutionary time, finer and finer means for dissembling, by acquiring greater neurological control over the vocal apparatus. Meanwhile, the listener, who has his own survival concerns, gets better at picking out the particular blend of pitch, rhythm, timbre and volume that marks the vocalizer as a deceiver. This compels the sender to further refine his "manipulations," which creates further pressure on the receiver to improve his acoustic "mindreading." What you end up with, after hundreds of millions of years of such coevolutionary one-upmanship, is an animal (us) who can detect, in just a few syllables, evidence of emotional panic in a father, betrayal in a spouse, lust in a dinner companion, or disapproval in a boss.

Scientific inquiry into precisely how the voice encodes such subtle emotions was, for all of the twentieth and much of the early twenty-first centuries, sporadic at best, nonexistent at worst. The paucity of research was mostly due to the nearly insuperable difficulties of investigating affective states in our species. Even labeling specific vocal emotions is difficult: what to one person sounds like fear can sound, to another, like anger. Is that "joy" or merely "enthusiasm"—or "excitement"? In the study of dog, bird, or squirrel monkey vocalizations, the problem is negligible, because of the ease in matching their expressive noises to the actions they take. But we humans are (thanks to our oversized cortex and the ways it interacts with our limbic system) far more *complicated*.

Imagine you are a voice researcher specializing in emotion. Even supposing you can faithfully recognize and label highly specific vocalizations, how do you make your human subject produce a real, true, *felt* emotional voice sound on cue? Say you want to study "anger." Or "sadness." Or "pride." How about "jealousy"? Never mind the truly fascinating, complicated, blended emotions like "Guilt-mixed-with-a-bit-of-passive-aggression." Even supposing that you manage somehow to elicit one perfect sample of a given vocalization (and can recognize it for what it is), how do you get your subject to reproduce that exact sound, over and over again? (Science demands repeatability, to make sure you're not getting a one-off anomaly.) Imagine trying to determine how such sounds are produced by the complex action of the various laryngeal muscles. Not easy, when your test subject is pin-cushioned with EMG needles—electrodes pushed through the neck into the larynx muscles to monitor the tensing and relaxation that control pitch. ("Give me a vocalization that says, 'Restful ease.'")

Given such difficulties, it is perhaps not surprising that very few scientists of the last century were willing to devote their career to the study of vocal emotion. The great exception is Klaus Scherer, a German-born social psychologist who is, at age seventy-five, the indisputable world authority on voice and emotion. He earned this accolade because he, almost alone among scientists, was stubborn and obsessive enough to master the wide array of disciplines (anatomy, physiology, neuroscience, phonetics, linguistics, auditory physics, psychology, computer programming) necessary to make a dent in the subject.

Born in 1943, near Cologne, Scherer traces his fascination with the subject to his teen years when he acquired one of the first commercially available tape recorders and began producing radio plays and documentaries with his high school friends. This, along with a love of the BBC radio soap opera *The Archers*, impressed upon Scherer how voices, even when isolated from all other social cues (facial expression, gestures, clothing), vividly conjure specific people, their personality, character, and emotions.[17]

He began formal study of the voice in 1968, when he enrolled in the PhD program in social psychology at Harvard. His initial interest was in voice and personality, how traits like "charisma" and "persuasiveness" are communicated by vocal acoustics. Scherer organized test subjects into mock juries to argue cases and measured their personalities, using peer ratings, to determine who was dominant, who was passive, and how that was manifest in the vocal signal. That these traits were mostly dependent on just two predictable variables (volume and pace) was both a surprise to Scherer—and a disappointment. He was in the midst of this research (which he found increasingly "boring") when he attended a talk by visiting lecturer Paul Ekman, a University of California, San Francisco, psychologist known for his pathbreaking study of facial expressions. In visits to a remote tribe in Papua New Guinea, Ekman documented how the tribe communicated basic emotions—happiness, anger, disgust, sadness—with expressions identical to people in London, Paris, New York, or anywhere else, supporting an idea first noted by Darwin in *The Expression of the Emotions in Man and Animals*.[18] Fascinated, Scherer approached Ekman after the lecture and mentioned his research on the voice. Ekman encouraged Scherer to drop personality research and focus on vocal *emotion*. "I'm the face man," Ekman told him, "but maybe you're the *voice* man."[19]

Scherer decided to become just that. After completing his PhD in 1970, he took a professorship at Giessen University and began designing experiments. Like Ekman, Scherer took a Darwinian approach, seeing the emotional channel of the voice as a mechanism, selected and perfected over millions of years, for promoting the survival and reproduction of our species, and thus a universal trait that should sound the "same," in terms of emotional prosody, no matter what language someone was speaking. He found early support for this in an experiment that used test subjects from Europe, Asia, and the United States who listened to a speaker expressing an array of states (fear, anger, joy, surprise, sadness). The listener-judges, despite not knowing the language, accurately identified the specific emotions 66 percent of the time: strong evidence that vocal emotions are, like facial expressions, universal. You actually know this from watching subtitled films where a Japanese or German or French actor's line readings "fit" exactly the emotional prosody of your native tongue—otherwise, what you see the actors doing, and hear them saying, would clash with the subtitles in a way that would make the movie unwatchable.

But when it came to *quantifying* specifics of the vocal signal that convey precise emotions, Scherer encountered challenges very different to those confronted by Ekman. Facial expressions can be photographed, filmed, and closely scrutinized, the muscle groups readily mapped and compared across individuals, cultures, and species. Not so the voice, whose signal results from an extraordinarily complex interaction between diverse body parts (most of them hidden inside the body) acting upon invisible air molecules. To further complicate matters, the emotion-bearing air vibrations also encode the speech signal (all those vowel formants and consonants that comprise words), and other elements like the pitch and pace variations that mark noun and verb phrases—that is, the linguistic prosody that helps to convey meaning. From this richly layered, complex signal, emotions must somehow be teased out.

Even harvesting data for research was fraught with difficulties. The most "ecologically valid" data were from recordings of real-life emotional situations, like cockpit recordings of pilots in life-threatening situations, live news reports of fatal disasters (like the *Hindenburg* crash where the radio announcer, in a live broadcast caught on tape, has a vocal meltdown as he witnesses passengers and ground crew being immolated), harassed callers to tech support lines, and even excitable contestants on TV game shows. But such data tends to be of short duration and poor sound quality (recorded through telephones, crackling cockpit PA systems, and small TV speakers). It also requires an investigator to rely on subjective inferences about what emotion a speaker is feeling. Does the radio announcer's cry of "Oh, the humanity!" as he watches the *Hindenburg* go up in flames indicate "sadness," "horror," "fear," or some other emotion—or a blend of all of them? Greater experimental control, Scherer found, is offered by "induction studies," in which stress reactions are produced in test subjects by exposing them to stimuli (like unsettling pictures or films), or asking them to complete stressful tasks, like guiding a helicopter avatar in a videogame under time pressure, while they vocalize. But induction studies yield relatively weak emotional responses (the pilot of an *actual* disabled airplane barreling toward a hangar makes different sounds than a gamer manipulating pixels), and, again, labeling the vocalizer's internal state can be problematic. "In spite of using the same procedure for all participants," Scherer wrote in a 2002 review paper, "one cannot necessarily assume that similar emotional states are produced in all individuals."[20]

The best method, Scherer concluded, is one that was used for the first-ever investigation into emotion in everyday speech, a study that used actors speaking scripted content. That experiment, conducted in 1931 at the University of Iowa, was by Gladys Lynch, a PhD candidate who wanted to identify the vocal parameters that distinguish good from bad acting. She recruited twenty-five trained thespians, and an equal number of nonactors, to read aloud prose passages with widely differing emotive content: a dull technical manual; a Galsworthy play about angry, striking factory workers; and a Seán O'Casey play in which a mother laments her child's death. Meticulous analysis of the pitch, pace, and volume of each recorded voice revealed that, for all the variations that the professional actors introduced into the lines to supply a certain theatrical liveliness and originality to their performances, *all* the voices followed an identical pattern of rises and falls—as if each written passage contained not only the literal meaning encoded in the words, but an emotional meaning encoded in the melody that any human voice, theatrically trained or not, will express.[21] The study strongly supported the Darwinian notion of a universal emotional grammar in our species. But critics argued that actors performing scripted content yield questionable data because they exaggerate obvious emotional cues and miss more subtle ones. Scherer, however, defended the use of actor-portrayal studies after his research of the mid-1970s showed that listener-judges, across languages and cultures, could accurately identify highly specific vocal emotions produced by actors—proof that many aspects of these simulated vocal signals were encoding emotions accurately.

But Scherer added a fascinating proviso. The actors in such studies would ideally have been trained in the performance technique developed by the nineteenth-century Russian acting teacher Konstantin Stanislavsky, later called the Method, and popularized by Hollywood stars including Marlon Brando, James Dean, Robert De Niro, and Meryl Streep. Such actors eschew the practice of classically trained thespians like Laurence Olivier, whose technique consisted of closely observing, then scrupulously mimicking, outward emotional expression.[22] Method actors recall (and relive) personal, emotionally salient events from their *own* lives and allow their internal reactions to take the spontaneous form that emerges in behavior. This can lead to breathtakingly genuine-seeming emotional expression, as is clear from Brando in *A Streetcar Named Desire*, De Niro in *Taxi Driver*, or Streep in *The Deer Hunter* (or any movie she's ever been in)—performances that revolutionized acting.

The neurological explanation for *why* the Method works (and thus its efficacy in emotion studies) can be found in research of the mid-nineteenth century by the French neurologist Guillaume Duchenne. In a study of the facial muscles, Duchenne showed that a genuine-looking smile involves two distinct muscle groups: one that lifts the corners of the mouth and a separate, crescent-shaped muscle that surrounds the eye socket which makes the outer corners of the eyelids crinkle in a highly specific way. Both muscle groups must be activated for the smile to look real. (Darwin, in his research for the *Expression of the Emotions*, tested this by showing to his house guests two photographs taken by Duchenne, one of a man smiling only with his mouth, the other of the same man smiling with both mouth *and* eyes; twenty-one out of twenty-four of Darwin's house guests correctly labeled the non-crinkly-eyed smile as fake.) Duchenne explained why the two smiles communicate such different messages. The muscles that pull up the corners of the mouth are under our conscious control—as is clear when the wedding photographer, for the fiftieth time that day, shouts, "Say *cheese*!" But the crescent-shaped muscles around our eyes aren't under our voluntary control (as those dead-eyed wedding photo smiles attest). These muscles are "only brought into play by a true feeling, by an agreeable emotion,"[23] Duchenne wrote.

That the muscles which give rise to emotional expression (including those of the voice) are beyond our voluntary control is a central tenet of the Method, which uses personal memories to circumvent the conscious mind, the cortex, to activate those parts of the emotional brain—the limbic system—which produce the involuntary movements that give rise to "real" emotional displays in face, gesture, posture, and voice.[24] That's how

Brando stunned audiences with a performance both tender and violent in the theatrical and movie versions of *A Streetcar Named Desire* (directed by Stanislavsky acolyte Elia Kazan), and De Niro galvanized movie audiences with his weirdly polite, yet ominous challenge ("You talkin' to me?") in *Taxi Driver*, or Streep brought tears to the eyes as the abandoning mother in *Kramer vs. Kramer*, when her voice trembles so realistically during her testimony in her divorce trial.

And that's why Scherer began using Method actors in experiments aimed at distilling, from speech, the parameters of pitch, volume, pace, and rhythm that convey specific emotions. In one of his most ambitious experiments, from the mid-1990s,[25] he recruited twelve actors (six men, six women) to portray fourteen finely discriminated affective states, including "hot" anger, "cold" anger, panic fear, anxiety, elation, boredom, shame, and contempt. To eliminate the contaminating influence of language (whose associations might color a listener's judgment of the emotion being voiced), Scherer and his co-researcher, Rainer Banse, constructed a pair of sentences that combined features of six European languages randomly arranged into seven-syllable nonsense utterances:

Hat sundig pron you venzy
Fee got laish jonkill gosterr

Scherer was careful not to instruct the actors according to explicitly labeled states (for example, "Read this as if you're very angry") since the label "angry" might be interpreted differently by each performer. Instead, he gave them scenarios (like "death of a loved one"). With each actor performing fourteen emotions, using the two nonsense utterances, in two takes each, Scherer and Banse generated a corpus of 672 voice samples, which they winnowed, according to various criteria (including sound quality) to 280; these were played to student listener-judges who made a final selection according to which portrayals best matched the intended emotion. Two emotions were immediately identified as almost impossible to hear from acoustic clues and eliminated from the study: shame and disgust. Scherer speculated that disgust is often expressed acoustically in a single vocal burst (for example, "yuck!") and thus sounds unfamiliar when spread out over seven syllables. Shame might have evolved few vocal cues because when people feel that emotion, they tend to clam up.

To analyze the remaining 224 actor portrayals, Scherer and Banse developed a custom software program that could parse, at a scale of milliseconds, minute shifts in pitch, volume, pace, and spectral acoustics. The pair detailed how high-arousal emotions (like "hot anger," "panic fear," and elation) showed a marked increase in both pitch and loudness. Sadness and boredom had a lower pitch and volume (since people in those depressed states can't summon much energy, either in breathing or vocal cord tightening, to produce anything but a low murmur). Contempt had a low pitch but also a low volume, a correlation that Scherer and Banse explained, in psychological terms, with a reference to Darwin's reflections on animal voices, in which "superiority displays" are expressed by a deepened pitch to suggest greater body size (and thus dominance). The "dampened volume," however, "may serve as a signal to the recipient that the sender . . . does not consider the other worthy of the expenditure of energy." The same lowered pitch *and* volume is used in sarcasm, like the teen "praising" his father's new khakis—a withering effect heightened by the contrast of the languidly blasé vocal tone with the surface praise in the words. ("They look *great*.")

The study offered an unprecedentedly exhaustive analysis; one chart, filling an entire page, showed the fourteen emotions broken down into fifty-eight separate acoustic variables, for a total of 812 measurements, each calculated to multiple decimal places. Exactly who might benefit from such fanatically microscopic analysis was of no concern to Scherer, who, as someone performing pure "basic science," has always believed that such knowledge is worthwhile *for its own sake.* It also perhaps explains why, for much of his forty-year career, Scherer has been an outlier, a lone traveler in an underpopulated, and largely unsung, branch of science.

But that changed in mid-2017, when the quest to understand vocal emotion became, overnight, a Holy Grail of science, thanks to a collective realization on the part of Silicon Valley that the Siri and Alexa speech recognition functions on our mobile devices will be infinitely enhanced when they can decipher, not only the words we speak, but the emotion in our voice, and to produce appropriate emotional, and thus human-sounding, speech in reply—like the disembodied computer voice that Joaquin Phoenix's character falls in love with in the movie *Her* (voiced by Scarlett Johansson), a voice that speaks with such fidelity to the prosody, paralinguistics, and timbre of the human vocal signal that it is impossible to tell apart from the real thing. Such astonishingly realistic computer-generated speech is still a fantasy, but not for long, if the top tech companies have anything to do with it. And they do. Which is why Scherer, after decades toiling almost alone, now finds his subject at the center of research programs driven by the richest companies in the world: Google, Apple, Amazon, Microsoft—all of whom are in a dead heat to develop the disruptive, game-changing voice-emotion software that will transform our relationship with our computers and which, some believe, will mark the next step in our evolution (or perhaps devolution) as a species.

The branch of science that seeks to imbue computers with emotions was first conceived of and named in 1995 when Rosalind Picard, a thirty-three-year-old assistant professor at MIT, published her landmark paper, "Affective Computing."[26] At a time when iPhones were not even a gleam in Steve Jobs's eye and the World Wide Web was accessed by dial-up modem, Picard laid out a vision for how computers could (and should) be equipped with the ability to detect and communicate emotions. "Most people were pretty uncomfortable with the idea," Picard (now head of MIT's bustling Affective Computing lab) recently recalled. "Emotion was still believed to be something that made us irrational . . . something that was undesirable in day-to-day functioning."[27] Picard made the case for affective computing by citing Damasio's research on the role that emotions play in shaping reason. "The neurological evidence indicates emotions are not a luxury," she wrote, "they are essential for rational human performance."[28] Indeed, Picard believed that full Artificial Intelligence, AI, would not be achievable until computers possessed not just the ability to crunch massive amounts of data (an act humans perform with our cortex), but to perceive and express emotion (which we do in our limbic brain).

The idea of "emotional computers" aroused skepticism, and even ridicule, several years into this century, according to Björn Schuller, a professor of Artificial Intelligence at Imperial College, London, and the cofounder of a computer voice-emotion start-up company that is, today, among the fastest growing in the world.[29] Schuller told me that he was a "born computer nerd" who first became interested in computerized vocal emotion at age nine, when he saw the American television series *Knight Rider* and its Artificially Intelligent talking car, KITT. "In the first episode," Schuller recalls, "KITT says to his human owner [played by David Hasselhoff], 'Since you're in a slightly irritated mood caused by fatigue . . .' I was, like, 'Wow, the car can *hear* that he's irritated and he's tired!' This totally got me."

When Schuller started his PhD in computer science at the Technical University of Munich in 2000, his focus was on *speech* recognition—the effort to transcribe spoken words and sentences into written text using computers. It's a harder task than it might sound, given the way we smear acoustic information across syllables according to how we move our mouths as we talk. The two *c*'s in the word "concave," for instance, are completely different sounds, because of how you round your lips in anticipation of the upcoming *o* when you say the first *c*, and how you retract your lips in anticipation of the *a* vowel for that second *c*. (Say the word slowly while looking in a mirror and you'll see what I mean.) We think they're the same sound, but computers know they're not. Which led to all sorts of absurd transcription errors in the early days of computerized voice-to-text transcription. Steven Pinker has noted some of these (including "A cruelly good M.C." for "I truly couldn't see," and "Back to work" for "Book tour"). To avoid such mistakes, programmers had to input every possible variation for how a *c* or a *d* or an *n* can sound in the various contexts in which it crops up in speech. The *n* in "noodle" but also "needle"—to say nothing of the *n* in "pan," or "pin." They had to do this for every consonant and vowel combination in the language—a Herculean, not to say Sisyphean, task that left some doubt as to whether any computer, anywhere, could ever accurately decipher the human voice and render it in text, to say nothing of the still harder task of convincingly reproducing it in simulated speech.

That all started to change around 2005 with the advent of machine learning, a new way to write software. Instead of laboriously inputting a bazillion speech sounds, software engineers started to write algorithms that teach a computer how to *teach itself* by listening to huge amounts of human speech, analyzing it, and storing in memory the particular way the individual sounds are pronounced when placed in particular contexts (or "coarticulated," as linguists put it). When Björn Schuller started his PhD in speech recognition, the dream of reliable text transcription by computer seemed decades in the future; but within a couple of years the coarticulation problem was more or less solved.[30] Schuller began to look for new horizons and, recalling his childhood fascination with KITT the talking car, he began to wonder if there was a way to make computer voices actually *sound* human—that is, by imbuing their speech with emotional prosody. Most people considered this a pipe dream—the fantasy outlined in Picard's theoretical "Affective Computing" paper of six years earlier. Nevertheless, when Schuller heard a fellow first-year PhD student complain that she had quixotically accepted the challenge from a local tech company to create a video system that could read the emotions in facial expressions, he was intrigued. "At the time," Schuller recalls, "computer vision systems were nowhere near good enough to detect emotions in the face. But I was studying speech, and I knew there was audio on my friend's video footage. So, remembering *Knight Rider*, I said to her, 'Give me your data. Let's see where we get.'"

Schuller wrote a program for detecting some basic changes in pitch, volume, and pace and used it to analyze the emotional content in the audio portion of his friend's video files. "It worked to some degree," he says, "and I got totally fascinated." Indeed, he instantly switched his PhD from speech recognition to speech *emotion* recognition. "All my colleagues made fun of me," he says. "They wouldn't take me seriously. Until maybe 2007, or 2008."

By then, computers had—thanks to increased processor speed and computing power—essentially mastered the coarticulation problem, eliminating all but a few of the errors Pinker documented. Big Tech—Google, Apple, Microsoft—were hungry for the *new* new thing and began to turn their attention to the missing ingredient in computer speech: emotion. By 2012, the landscape had changed completely. "The focus of research," Schuller says, "shifted totally toward what *I* do."

What Schuller, and a growing number of others, "do" is use machine learning to make computers teach *themselves* emotional prosody. Schuller plays accurately labeled samples of emotional vocalizations into the computer's learning software ("angry," "sad," "happy"), and the machine does the rest.[31] And the algorithms are learning fast. At present, computers can recognize specific vocal emotions 65 to 70 percent of the time, about the same as humans—astonishing progress given that the field is less than a decade old.

For pioneering voice-emotion researcher Klaus Scherer, who spent a half century teasing out the measurements in the various elements of the vocal acoustic signal, these developments are bittersweet. He is gratified by the sudden widespread interest in a subject to which he devoted his life, but disheartened that today's computer engineers take not the slightest interest in the voice signal's quantitative measurements—all those minutely calibrated tables of numbers and values for pitch, volume, rhythm, pace, and overtone frequencies that Scherer painstakingly accumulated from his experiments with Method actors. Computer engineers like Björn Schuller don't have to know anything about the minuscule acoustic adjustments that distinguish anger from fear, or joy from irritation: they need only play recordings of properly labeled voice emotions into the computer's learning software. The results are amazing. "But," Scherer adds, with some bitterness, "it doesn't really mean anything in terms of our understanding of how it all *works*."

Scherer is also frustrated that speech emotion engineers study only a handful of "basic" emotions—fear, anger, joy, sadness, boredom, surprise. Complex, nuanced, blended emotions are considered too difficult at present to label and simulate. Scherer doubts that any technology will ever be able to decode the most complex vocal emotions—like the panic underneath my "cheerful" welcome to my son during the global fiscal collapse, or the hint of marital betrayal in the husband's request for the remote, or the threat-sound we pick up in De Niro's voice when he asks, "You talkin' to me?" "That kind of voice detection comes about through a *channel discrepancy*," Scherer told me. "There is something falling out between the voice quality on one hand, and the prosody on the other. The two don't jibe." From the resulting discordance, a listener draws psychological inferences.

"When you say, cheerfully, 'Hi!' and your son responds '*What's wrong?*' he hasn't detected *what's* wrong," Scherer went on, "but he has detected that *something* is wrong. Same with the suspicious wife. She hasn't heard a vocal cue that specifically encodes *guilt* or *sexual betrayal*, but she's heard a discrepancy. What she comes out with—*Are you having an affair?*—is an expression of the underlying prejudices and fears that she has. That we *all* have. With a discrepancy, we project our innermost fears onto that thing." Scherer believes that computers will never be able to grok the incredibly complex interplay between acoustic analysis, psychological deduction, and emotional projection inherent in such acts of hearing. "I think the human ability to make *conjectures* on the basis of vocal cues is unique," he told me, "and I think we'll be able to keep that advantage over computers."

Björn Schuller disagrees. He says that the "conjectures" we draw from such channel discrepancies are possible because the listener is *so familiar* with the speaker's particular voice: a father and child; a husband and wife; a boss and employee (or what we know of Travis Bickel's antisocial personality from the movie's opening scenes). Such familiarity is already being established between us and our iPhones and Androids, Siris and Alexas. The preinstalled speech-learning software in our devices is constantly educating itself about the idiosyncratic way we pronounce our vowels and consonants (to better "understand" what we're saying). But when Google, Apple, and Microsoft start installing *emotion-based* machine learning software on our devices, as Schuller expects them to do in the next three to five years, the learning curve about other particulars of our uniquely expressive voices will spike dramatically and in dimensions far beyond our specific way of coarticulating *c* before short *o*. Our Alexas and Siris will analyze voice dimensions like *arousal*, which, in emotion studies, is defined by how calming versus how exciting the signal is; *valence*, which refers to how positive versus how negative is the feeling expressed; and *dominance*, which measures the degree of control versus the degree of submission communicated by the signal. In this manner, your computer will know the emotional makeup of your specific voice as well as your mother does. You will be as incapable of disguising your true feelings from your iPhone or Android as you are from her.

Or so Schuller hopes. In 2018, he began working with MIT's Rosalind Picard, the original oracle of affective computing, on a technique they call "personalized machine learning," designed to teach your computer the nuances of your voice so well that the false cheerfulness in the question "Hi, how was school?" will instantly be detected. Humans do this by synthesizing and analyzing, at unimaginable speeds, the vast amounts of acoustic data in the voices we know. He and Picard believe that machine learning will give computers the same power. "Then we will be at this point where you say to your laptop, 'Yeah, everything's perfect, computer!' And the computer says"—Schuller adopts a tone of sarcastic skepticism—"'Yeah, *sure* . . .'" He cites as possible applications for such software the early diagnosis of autism, detecting depression or suicidality in teens, and catching mental illnesses in workers before they act out in mass shootings or other antisocial behavior.

But there is a potential cost to computers that, in their emotional awareness, can masquerade as humans. The proliferation of computer "bots" on Twitter and Facebook helped to drive the Brexit vote and Trump's election as president—and it is for this reason that one tech critic called emotional voice simulation a "Pandora's box"—yet another way that political operatives, and other mischief makers, will be able to construct false realities undetectable to our human sensory apparatus. Schuller says that he began hearing such criticism after the electoral shocks of 2016. "For the first time ever as a researcher," he told me, "I've been getting negative feedback. In the past, people were always, like, 'Oh wow, that's cool!'"

Picard has never been blind to the potential dangers of affective computing. In her seminal 1995 paper, she cited as a warning Stanley Kubrick's *2001: A Space Odyssey*, in which HAL, the super-intelligent computer that runs the Jupiter-bound *Discovery* spaceship, suffers an *emotional* meltdown, triggered by fears that the crew members plan to shut "him" down. HAL kills all but one of the astronauts. "The message is serious," Picard wrote: "a computer that can express itself emotionally will someday act emotionally"—a result, she says, of its "mimicking both cortical and limbic functions."[32] That computers control "the phone system, stock markets, nuclear power plants, and jet landings" (to say nothing of thermonuclear launch codes) is, Picard concedes, cause for concern. Nevertheless, she, like Schuller, believes that the benefits of emotional computing[33] will outweigh any potential risks—even if those risks include, as Picard puts it, that "we, the maker, may be eliminated by our creation."[34] Whether this is an unreasonable fear is, of course, impossible to say at this point. What is not in doubt, however, is that much of the most cutting-edge work in imbuing computers with emotion starts with machine mastery of our infinitely subtle, affectively expressive, prosodically nuanced, paralinguistically rich human voice.

FOUR

LANGUAGE

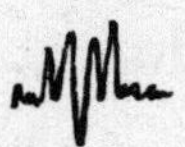

For now, anyway, we remain the only entities, animal or machine, capable of blending emotion and language in a single vocal sound wave. How we developed this remarkable and, so far, inimitable talent is unknown. The brain, lungs, larynx, tongue, and lips do not fossilize, so we can't sift through the remains of extinct species for clues to how the cries and calls of our animal ancestors became speech, and the sounds themselves have, of course, long since fled from the air. Which is why the riddle of language origins has been called the "hardest problem in science."[1] It has nonetheless obsessed humanity since antiquity, given rise to some of the bloodiest and most fascinating intellectual battles ever fought, and provided remarkable insight into who we are and how we came to be.

Modern theories of how speech evolved date to the Enlightenment, when Johann Gottfried Herder, a student of Immanuel Kant, published his *Treatise on the Origin of Language* (1772) and asserted that words began as an onomatopoeia-like imitation of natural sounds using the voice: bleating of sheep, cooing of turtledoves, barking of dogs, rustling of leaves, rippling of water, whooshing of wind. Vocal mimicry of these sounds formed a pre-language capability, a "proto-language"—the crucial leap from howls and shrieks and chirps to *articulated* sounds that refer to things in the world (the thing my parakeet, and no other nonhuman animal, can do). Actual language, with fully formed words and syntax, gradually developed from there. Herder's insights are all the more remarkable considering that he advanced them a century before Darwin informed us that we once existed in a more primitive state from which our intellectual faculties, including language, slowly evolved.

Other Enlightenment thinkers offered competing theories—including the French philosopher Étienne Bonnot de Condillac, who said that words developed not from imitating sounds in the outside world, but from impulses *within*: interjections, like the *Arrgh!* we shout upon hitting our thumb, or the *Oooo* we murmur at a pleasurable touch.[2] (Herder, predictably, said this was insane.) Still others said language arose from the grunting that accompanied hand gestures, or the noises of sympathy we use to encourage or comfort others. The debate over the origins of speech grew so vicious and childish over the next hundred years that a leading research institute, the Société de Linguistique de Paris, in 1866 banned all further discussion of the topic.

A major catalyst for this prohibition was a famous address by the era's leading linguist, Oxford University professor Max Müller, who, in 1861, delivered a series of lectures that attacked all existing theories of language's origins. He ridiculed Herder's ideas about onomatopoeic mimicry, calling it the "Bow-wow theory," and dismissed Condillac's ideas as the "Pooh-pooh theory."[3] Müller had a hidden agenda in these attacks: a devout Christian, he adhered to the Bible's explanation that language began when God, as recounted in the book of Genesis, invited Adam to name all the animals.[4] As such, Herder and Condillac were mere collateral damage in Müller's campaign against a far larger target: Darwin, whose *Origin of Species* had been published eighteen months earlier.

Müller correctly viewed the *Origin* as the single greatest threat to religion ever disseminated, but he also believed that he had spied the book's Achilles' heel. Darwin had deliberately not discussed *human* evolution, leaving it up to readers to infer how our exceptional *mental* powers, the most conspicuous of which is language, arose. Even if natural selection could account for the complexity that creates an eye or hand or lung, Müller said, it could never explain something as intricate as speech. "Language is our Rubicon," as he put it, "and no brute will dare to cross it . . . no process of natural selection will ever distill significant words out of the notes of birds or the cries of beasts."[5] In short, Müller became the standard-bearer for the view that speech is entirely discontinuous from animal voices.

Müller saw his argument taken up by creationists and other opponents of Darwin, including (amazingly) the co-discoverer of natural selection, Alfred Russel Wallace, who, having veered into spiritualism later in life, declared that, when it came to language, some mysterious "higher intelligence" had to have intervened.[6] All of this meant that, by the time Darwin finally sat down to address *human* origins in his second major book on evolution, *The Descent of Man*, which he began writing soon after Müller threw down his gauntlet, the question of language origins loomed large—indeed, as the single most significant riddle Darwin had to solve.

In his efforts to convince readers that articulate humans descended from hooting chimpanzees, Darwin could not even offer concrete evidence of our ancestral descent from simians—since no fossil remains of linking species between apes and humans had yet been discovered. After Darwin's death (in 1882), scores of such fossils were found: skeletal remains of nearly two dozen hominid and hominin species, tracing a clear line of descent, over fourteen million years, from the apes of eastern Africa to Homo sapiens.

Lacking that evidence, Darwin was obliged to use the comparative method, which meant pointing out the telling similarities, and differences, between living apes and us. He argued that the transition from simian knuckle walking to our upright posture freed the hands for tool making and other digital manipulations, which led, through positive neural feedback, to our enlarged brain and increased mental powers.[7] At the same time, Darwin was at pains to stress how our mental faculties differ only in *degree*, not in kind, from other animals, who also possess "higher mental powers," including memories, reasoning, tool use, even "architecture and dress" (in the shelters baboons build and the vegetation they drape on their bodies when cold). In thus arguing for how surprisingly *small* is the mental gap between nonspeaking animals and articulate humans, Darwin sought to reduce Müller's "unbridgeable" Rubicon River to a mere trickle. He was then ready, in a short chapter entitled "Language," to explain how our ape-human ancestors traversed that insignificant barrier—and started speaking.

Darwin begins by conceding that language "is peculiar to man," although not speech. Parrots, he reminds us, talk—although not to convey complex meaning. Only humans possess a "large power of connecting definite sounds with definite ideas"—a result of our unprecedentedly large, powerful brain. But intelligence is only part of the story, he said. Language is, like animal sounds, "instinctual," as is clear from the automatic babbling of babies. Still, speech cannot be a "true instinct" because "every language has to be learned" through exposure to adult speech. In this, Darwin said, we are just like birds—one of the only other species who acquire their specific vocalizations from listening to their parents. (Dogs, cats, horses, chimps, indeed every mammal, are born with their vocal signals hardwired; they will make their species-specific noises even if they've never heard such sounds.) Furthermore, birds, during the vocal learning stage, take on the local "provincial dialect" of the surrounding adult birds—just as human babies, displaced from their birth location, acquire the language, and regional accent, of their adoptive home.

These uncanny correspondences between birdsong and human speech led Darwin to a highly original insight. Whereas all earlier theorists imagined *words* coming first, Darwin said that the melody and rhythm of speech, its birdsong-like pitch sequences across sentences—its *emotional prosody*—preceded words in some now extinct singing ape-human. He cited the living example of gibbons, a species of singing ape whose operatic mating calls are "true musical cadences . . . serving to express various emotions, as love, jealousy, triumph, and serving as a challenge to . . . rivals."[8] From this complex vocal music, he mused, speech gradually emerged. Which explains why Darwin was so surprised, several years after writing *The Descent of Man*, to discover, among his papers, the forgotten "baby diary" of his son William, which documented an identical sequence of verbal development: musical voicings like the upward pitch rise on the nonsense syllable *mum*, a melodic vocalization that gave way, eventually, to the articulated request for "food"; it seemed that William's speech had developed precisely as Darwin imagined speech developing in our species—emotionally expressive musical phrases giving way, in some slightly more mentally advanced ape-human, to a controlled movement of the articulators and the first wordlike sounds of a protolanguage.

In describing this protolanguage's emergence, Darwin imagined how our distant ancestor might have mimicked other species' voices: "It does not appear altogether incredible that some unusually wise ape-like animal should have thought of imitating the growl of a beast of prey, so as to indicate to his fellow monkeys the nature of the expected danger."[9] Darwin also embraced, without apology, the explanations (ridiculed by Müller) that were offered by Enlightenment thinkers of a century earlier. "I cannot doubt," he wrote, "that language owes its origin to the imitation and modification, aided by signs and gestures, of various natural sounds, the voices of other animals, and man's own instinctive cries."[10]

Once this protolanguage was in place, habitual use of the voice "strengthened and perfected" the vocal organs, even as it "reacted on the mind by enabling and encouraging it to carry on long trains of thought"—like those similar to the mental actions we use with numerical figures for algebra. From these linguistic "calculations" emerged grammar and syntax—spoken language, the embodiment of thought. To establish this link between thought and speech, Darwin drew on the most up-to-the-minute research on the left hemisphere brain center for language discovered by Paul Broca, in stroke patients, just a few years earlier.[11]

Despite its elegance and economy (some say sketchiness), Darwin's explanation did not settle the debate. The *Descent* was published in 1871. A year later, the London Philological Society joined the Paris group in banning all further papers on language origins. The topic vanished from serious scientific discussion for the next thirty years, until the dawn of the twentieth century, when Edward Sapir, a twenty-three-year-old Columbia University graduate student (who would go on to become one of the founders of modern linguistics), revived it. He did so when, for his master's thesis, in 1905, he wrote about the Enlightenment scholar Johann Gottfried Herder and his essay on the onomatopoeic origins of words. Sapir marveled at the "epoch-making" brilliance of a thinker who, a century before Darwin, "[did] away with the conception of divine interference" in language and replaced it with "the idea of slow . . . development from rude beginnings."[12] Sapir called for reviving the moribund study of language origins and wrote: "the path for future work lies in the direction pointed out by evolution." He recommended "the careful and scientific study of sound-reflexes in higher animals" and an "extended study of *all* the various existing stocks of languages."[13]

In writing those words, Sapir seemed to assume that some languages (specifically, those of indigenous tribes) are more "primitive"—*less evolved*—than those of "civilized" societies, and might contain clues about how speech emerged from animal vocalizations. But when he began his PhD at Columbia, he took a course taught by Franz Boas, the pioneering anthropologist whose paradigm-shifting study of indigenous peoples showed that all humans are created equal—that is, exist at the same state of evolutionary "advancement," and that any observable differences between, say, a city-dwelling investment banker and an igloo-dwelling whale hunter are attributable solely to *culture*, the customs and rituals that emerged to fit the particular environment in which the people live. To find out if the same was true of language, Sapir invented a new specialty: anthropological linguistics, and spent the summers of 1905 and 1906 doing fieldwork among various Native American tribes of Oregon. Through meticulous dissection of their speech—the sounds, vocabularies, and grammars—Sapir confirmed that no language is "simpler," "more primitive," or "less evolved" than any other. All partake of the same extraordinary, and mysterious, process of converting abstract thought into elaborately patterned acoustic signals with the voice. As Sapir put it: "the lowliest South African Bushman speaks in the forms of a rich symbolic system that is in essence perfectly comparable to the speech of the cultivated Frenchman."[14]

This prompted Sapir to shift focus from language origins to the question of how languages reflect specific cultures. In his classic book *Language* (1921), he urged others to follow his cultural approach. Generations of linguists duly set off, notebooks in hand, to collect and analyze the multifarious vocal sounds, vocabularies, and grammars produced by the inhabitants of the farthest-flung jungles, savannahs, plains, villages, towns, and cities, with an ear to figuring out why different languages sound as they do.

Sapir's cultural approach left untouched a host of questions about language, including how it is generated in the brain, how it is acquired in childhood, and how it got into our skulls in the first place. The first serious attempt to address these enigmas in the twentieth century was by the famous behaviorist B. F. Skinner.

Behaviorism's central tenet is that the brain, at birth, is a blank slate, and that all behavior is learned. The theory grew out of work by Ivan Pavlov, who taught dogs to salivate at the sound of a ringing bell, a form of associative learning called "classical conditioning." (Pavlov even taught dogs *not* to salivate at the sight and smell of food—arguably a more amazing feat.) Later behaviorists developed "operant conditioning," the shaping of behaviors through punishment and reward—like the lab rat, confined to a box, who learns to push a lever a precise number of times to earn a food pellet. Skinner became the world's leading behaviorist when he showed how such conditioned behaviors could be linked into long chains to create stunningly complex acts. A *New York Times* article from 1950 chronicled Skinner's success in conditioning chickens to play ping-pong and peck out the tune "Take Me Out to the Ball Game" on a keyboard.[15]

In his 1957 book, *Verbal Behavior*, Skinner said that speech acquisition in infants is also reducible to chains of operant- and classically conditioned responses inscribed on the blank slate of the brain after a baby is born.[16] We now know, from the remarkable research on speech acquisition detailed in this book's first chapter, that Skinner's theory was wrong in its details, but not in its overall argument that language can only be acquired through interaction with the environment. Skinner's description of that interaction was, to be charitable, eccentric: he imagined that the process of "conditioned" learning of language begins with the reward of parents' joyful reaction to the infant's first mush-mouthed phonemes ("I think he's trying to talk—*grab the video camera!*"), progresses to the parents' emphatically positive (or negative) reactions to first words ("Yes, Susie, that's a *dog*—very good!" or "No, Jake, that's a *cat*, not a *mat*"), and eventually culminates in the teaching of syntax by subtle messages of reward and punishment that reinforce the child's correct sequencing of words across sentences. Because this conditioning uses the identical punishment-reward process as lever-pushing in rats or tune pecking in chickens, the evolutionary roots of language were, in Skinner's view, self-evident.[17]

Skinner's explanation of language acquisition was characterized by its extreme "nurturist" stance, a view that speech is all learning and *nothing* about our ability to speak is inborn—a position in direct opposition to the one adopted, right around the same time, by Noam Chomsky. Indeed, it was Chomsky's demolishing 1959 review, in the leading journal *Language,* of Skinner's *Verbal Behavior* (he called it "empty," "vacuous," "false," "meaningless," and "just a kind of play-acting at science")[18] that put Chomsky (then an obscure thirty-two-year-old linguist at MIT) on the map, and which brought to the attention of the scientific world at large his view that language is not learned *at all*, but "grows like any other body organ." The theory would make Chomsky one of the most famous scientists in the world.

On the question of how language evolved, however, Chomsky was surprisingly incurious. Moreover, he shocked his scientific colleagues when he rejected the idea that language emerged through Darwinian natural selection.[19] When pressed for an alternate explanation, he mused that language somehow *appeared* through some unexplained physiological fluke resulting from (as he once put it) packing so many billions of neurons into a space "the size of a basketball"[20]—an explanation only marginally more creditable than the one offered in Kubrick's *2001*: the visitation by the mysterious monolith and the great leap in intelligence that results. Still more surprisingly, Chomsky insisted that, however language appeared, it did so not for sharing ideas, plans, and goals through vocal *communication*. Language first appeared, Chomsky said, for the purpose of solitary, pristine, platonic—and perfectly silent—*thought*.[21] Which ruled out, of course, any role, in the evolution of language, for the human voice.

Even some of Chomsky's greatest champions have acknowledged that he has become a "guru" in the field of linguistics, with all the attendant dangers that go with slavish adherence to a single person's ideas.[22] Not every linguist, however, followed Chomsky's dogma. The most notable apostate is Philip Lieberman, now an emeritus Professor of Linguistics at Brown University. For over fifty years, Lieberman has built his own theories of how language is learned and where it came from. In doing so, he has, virtually alone among serious scientists, proposed a central role for the voice.

Lieberman was one of the first students to enroll in Chomsky's linguistics course at MIT, in the late 1950s, before Chomsky published his evisceration of Skinner's *Verbal Behavior* and became famous. The class had just three other students.[23] Lieberman was at first enthralled by Chomsky's ideas, but soon came to question his conception of language as a purely *mental* phenomenon, a medium for thought. Clearly, Lieberman reasoned, language has an important *physical* dimension, as a *spoken* medium produced by the vocal organs for communication. Having earlier trained at MIT as an electrical engineer, Lieberman was, by temperament and training, fascinated by the physical apparatus that makes any complex mechanism function. Language was certainly a complex mechanism—as was speech, the primary means by which we transmit language. Lieberman wondered if clues to the nature and origins of language might not lie in the apparatus that makes language so useful to us: the voice.

For his PhD, Lieberman focused on the primary anatomical system that drives voice and speech—breathing—and performed groundbreaking work on an amazing, if underappreciated, aspect of our talent for talking: namely, how we gear the amount of air we inhale to the size of the thought we intend to speak, a lightning-fast calculation we perform depending on whether we plan to say "Pass the salt, please," or "Four score and seven years ago our fathers brought forth on this continent, a new nation, conceived in Liberty, and dedicated to the proposition that all men are created equal."This skill actually takes until puberty to master, as Lieberman showed by pointing to the speech of children as old as twelve who, despite clear articulation and considerable syntactic sophistication, sometimes find themselves running out of air midsentence.

Consulting respiratory physiologists, Lieberman learned how stretch receptor nerves in the rib cage's intercostal muscles—the muscles between each rib that expand the rib cage when we inhale, and compress it when we exhale—help us gauge when the lungs have been inflated to the "correct" amount for a given utterance. (Similar stretch receptors in the bladder notify us that it's time to find a bathroom; those in the stomach alert us to when we've eaten enough.) These stretch receptor nerves send signals from the muscles to the brain, but also in the opposite direction: from the brain to the muscles, creating a two-way circuit. Lieberman argued that this feedback loop in the nerves that control respiration, over the course of evolution, helped early humans to refine their breathing to fit the meanings they wanted to convey. We divide our thoughts into utterances that can comfortably be spoken on a single exhalation—which suggested, to Lieberman, an anatomical basis for one of the primary syntactic units: the sentence. We can even time our inhalations, he pointed out, around natural pauses *within* sentences, the places where, in writing, we would use commas to enclose an embedded clause (". . . a new nation"—*inhale*—"conceived in Liberty . . ."). The fact that we shape grammatical structures around our breathing seemed to Lieberman clear evidence of how an "anatomical constraint" (the amount of air we can pull into our lungs) helped to create the structures of language.

This put Lieberman sharply at odds with Chomsky, who said that our uniquely human, grammar-generating "language organ" developed purely as a medium for thought, and is sequestered in the highest regions of the cortex. But breathing is controlled by older parts of the brain—such as the brainstem and cerebellum, structures we share with every vertebrate. Perhaps, Lieberman thought, grammar and syntax are *not* confined to highly specific areas of the cortex unique to human beings—a dedicated "language organ." Perhaps it made more sense to think of language as distributed throughout the brain, in regions we share with our oldest nonhuman ancestors. In which case, Chomsky was surely wrong to say language could not have evolved through the normal processes of natural selection.

After graduating from MIT in 1966, Lieberman was hired as a full professor at the University of Connecticut and a member of the research staff at New York City's Haskins Laboratories, the leading language research institution that was shortly to publish the seminal motor theory of speech perception. At Haskins, he continued his investigations into the bodily changes that might have driven the evolution of language. One logical place to look for the origins of speech, he thought, was in the sounds and vocal apparatus of our closest living animal relations: apes.

At zoos in Manhattan and the Bronx, Lieberman taped hundreds of hours of chimpanzee, rhesus monkey, and gorilla vocalizations and concluded that, while apes make some good consonants, especially lip sounds, like *p* and *b*, and the nasals, *m* and *n*, their vowels were limited to the schwa, the "uh" sound[24]—a limitation that he realized, through his familiarity with research from the 1950s on how we form vowels, was imposed by their high-placed larynx. Consulting Victor Negus's classic *Mechanism of the Larynx*, Lieberman furthermore learned that we are the only animal with a permanently descended larynx, the result of the slow migration of the voice box down the throat that probably began when our ape ancestors stopped knuckle walking and stood upright. [25]

Our weirdly low larynx had actually puzzled Darwin, because it brings the opening to our windpipe right alongside the opening of our esophagus, the tube that leads to the stomach—a dangerous arrangement, Darwin noted, because of "the strange fact that every particle of food and drink which we swallow has to pass over the orifice of the trachea, with some risk of falling into the lungs."[26] In fact, until the advent of the Heimlich maneuver in the 1970s, thousands of people died annually by choking on food that "went down the wrong way." Darwin was understandably mystified as to why we evolved such an anatomical booby trap in defiance of everything he knew about natural selection, which is supposed to enhance our chances of living, not the reverse. Lieberman (in possession of knowledge that had not yet emerged in Darwin's day about how vowel sounds are created by the blending of the overtone frequencies produced in both the mouth *and the throat*) concluded that the descent of the larynx conferred an important survival advantage that outweighed the very real danger of choking to death: a throat resonating chamber that allowed for articulate speech through a variety of clearly discernible vowels.

It thus occurred to Lieberman that if he could track, in the fossil record, the descent of the larynx from its high position in the chimp's throat, to its uniquely low position in *Homo sapiens*, he might be able to create a rough timeline for when language emerged. The first hominid he studied was our closest extinct human relative: the Neanderthal, a human species that emerged around 500,000 years ago and which migrated out of Africa at least 200,000 years ago, populating large swaths of Europe and western Asia, before mysteriously going extinct about thirty thousand years ago—not long after encountering our species, *Homo sapiens* (who had eventually followed them out of Africa). Debate had long raged over whether Neanderthals could talk. They certainly seemed to have the necessary cognitive power (skull fossils show that their brain was slightly *larger* than ours). But some influential scientists insisted that language did not arrive until long after the appearance of Neanderthals—indeed, as recently as fifty thousand years ago, with the unexplained surge in cognitive power known as the Great Leap Forward. Chomsky endorsed the Great Leap theory. Lieberman didn't buy it. He was increasingly convinced that speech and language were gradual developments that depended on the coevolution of many different voice-related anatomical and cognitive systems stretching back hundreds of thousands of years—or hundreds of millions, if you went all the way back to the lungfish (or indeed the ragworm). In keeping with this, he strongly suspected that Neanderthals could speak, and that the placement of their larynx might offer strong proof of it.

Using a cast of a Neanderthal skull discovered, during the early twentieth century, in the village of La Chapelle-aux-Saints, in France, and working with Edmund Crelin, the chief of anatomy at Yale New Haven Medical School, Lieberman determined that the Neanderthal larynx occupied an "intermediary" position between the chimp and human larynx. In other words, it had begun to migrate down the throat but had not descended to the level of ours. Crelin happened to be the first anatomist in the world to have investigated the anatomy of human newborns and had earlier revealed the chimplike high placement of the newborn larynx.[27]

Over the first few years of life, the larynx gradually descends. That the identical downward journey of the larynx occurred in our early human ancestors was wholly in keeping with an idea in evolutionary biology advanced by Darwin: namely, that "ontogeny recapitulates phylogeny"—or, in plain English, that the stages through which humans pass, from conception, as a single-celled zygote in the womb, to maturity, mirror the developmental steps our species took from single-celled eukaryote, in the primordial ocean, to our evolution, 2.7 billion years later, as *Homo sapiens*.

After all, as fetuses we are suspiciously fishlike aquatic creatures who use gill-like structures to extract oxygen from the liquid in our mother's womb, and at birth we become air-breathing animals. In terms of our vocal tract, we are, effectively, chimpanzees who only gradually "evolve" into speech-capable humans as the larynx repeats the migration it made, after we diverged from Neandethals, almost to the bottom of the neck. Meanwhile, the sequential development of human vocal sounds parallels that of vertebrate brain evolution: a newborn's first cries emerge from the (reptilian) brainstem as purely reflex reactions to pain or hunger; emotionally nuanced sounds appear around the third month as the baby's mammalian limbic structures come online and its musical, prosodically expressive sounds help it forge social bonds with caregivers; only after that, when the larynx has descended sufficiently to allow the sculpting of rough vowel sounds, does the rational, executive cortex start to wire in, and the child begins to shape its emotional cries and whimpers, giggles and sighs, into the babbled phonemes and proto-linguistic "words" (like William's *mum*) that the infant will eventually mold into articulate human speech.[28]

For Lieberman and Crelin, the burning question was how did the Neanderthal's slightly higher larynx affect its voice? Crelin built elaborate reconstructions of the Neanderthal vocal tract from silicone, and they showed, using a computer program to model the vowels that the reconstructed Neanderthal airway could produce, that our closest extinct relative, owing to its elevated larynx, could produce only a restricted range of vowels, wider than the chimp's schwa, but not including our *ee, ahh,* and *oo* sounds—those requiring the most drastic reshaping of the two vocal tract resonators with the tongue.[29] One year later, MIT speech researcher Kenneth Stevens showed that the ideal proportions of throat resonator to mouth resonator, for clear vowels, is a 1:1 ratio, with the length of the mouth equal in length to the vertical section of the throat.[30] These "ideal" proportions did not occur until the apelike snout of the Neanderthal face flattened in our species. For these purely anatomical reasons, Lieberman said, Neanderthal vowels would have been blurred, ambiguous—a handicap that made all the difference to the respective fates of Neanderthals and *Homo sapiens*.

That our two species were sufficiently close, biologically and socially, to have mated and reproduced is a staggering fact only recently revealed by DNA recovered, and sequenced, from Neanderthal bones. As it turns out, every human being whose lineage can be traced to Europe or Asia (where *Homo sapiens* and Neanderthals briefly coexisted) carries 1 to 4 percent Neanderthal DNA; people whose direct ancestry is African carry almost none[31] (for the simple reason that *Homo sapiens* and Neanderthals didn't meet up—and *hook* up—until both had left that continent). So, we are, in *every* sense of the term, closer to Neanderthals than anyone had ever imagined. But it was this very closeness that, according to the merciless laws of natural selection, put our two species on a collision course. Lieberman cited Darwin's claim that "each new variety or species, during the process of its formation, will generally press hardest on its nearest kindred, and will tend to exterminate them."[32] Lieberman concluded that Neanderthal extinction "was due to the competition of modern human beings who were better adapted for speech and language."[33]

That is, we had better vowels.

In 1984, Lieberman published *The Biology and Evolution of Language*, the first book-length study on language origins since the Paris ban of 1861. It included his research on respiration, the descent of the larynx, vowel production, Neanderthal extinction, and other evidence to argue that language emerged, roughly 400,000 to 200,000 years ago, through naturally selected improvements in the vocal apparatus, which fed back into the brain, creating the complex neural circuitry (including the "language centers" of Broca's and Wernicke's areas) that make speech possible. He denied the existence of an innate "language organ" and attributed our linguistic capacity to our species' remarkable *general* intelligence—the same smarts that made us discover fire, build sophisticated tools, make protective clothing, and cooperate in large groups. The phenomenal speed of language acquisition in children was owing not to an inborn language instinct (Chomsky's Universal Grammar) but to Darwin's "instinct to *learn*."

Lieberman's book was, by any measure, the strongest challenge yet mounted to Chomsky, but in a field dominated by the theory of linguistic innateness and Universal Grammar, it was almost completely overlooked by the linguistic mainstream. Lieberman nevertheless continued his investigations into how changes to the speech apparatus gave rise to language, including research into an area that, because of its purely speculative nature, he touched on only lightly in *The Biology and Evolution of Language*: that our linguistic capability could only have come to full fruition (after refinements to our breathing and the descent of the larynx) with evolutionary changes to those parts of the nervous system that make possible the incredibly rapid, exquisitely well-coordinated, and meticulously sequenced movements of our tongue, lips, and larynx for speech. (To produce even as seemingly simple a sentence as "Pass the salt, please" requires the coordination and careful sequencing, in under one second, of hundreds of muscles throughout the trunk, larynx, tongue, lips and face.) Lieberman had included some preliminary findings about this in his book,[34] but few at the time were willing to credit the lowly *motor* system (parts of which we share with earthworms and mollusks) with a major role in the evolution of a higher-order process like language, which is, after all, an expression of thought. What possible connection could there be between *thinking* and the parts of the brain that control bodily *movement*? But since then, researchers have pinpointed cell death in the brain's motor centers as the cause, not only of Parkinson's patients' tremors and speech impairments (they mix up voiced and unvoiced sounds), but of disturbances in *thinking* known as Parkinson's Dementia. Lieberman's own study, in the early 1990s, of mountain climbers suffering from the temporary dementia of high-altitude sickness from oxygen starvation of the brain also showed the link between higher thinking, speech, and crucial movement centers in the brain.[35] The most dramatic evidence, however, was yet to come.

Many of the most momentous breakthroughs in science begin with the study of anomalies and abnormalities of human behavior—and such was the case for the evidence that began to emerge, in the late 1990s, in support of Lieberman's intuitions about the critical role, for language, in evolutionary changes in the motor control of our larynx, lips, and tongue. The anomaly, in this instance, was an unfortunate West London family with an extremely rare speech disorder that afflicted several family members over three generations (mother, children, and grandchildren) and included an inability to pluralize words by adding *s*, and confusion over how to make verbs into the past tense by adding *–ed*.[36] This *runs-in-the-family*, genetically determined disorder was initially seen as clinching evidence that our grammar and syntax must indeed be inborn, just as

Chomsky said, and that every aspect of language might be controlled by specific *genes* that build particular parts of the cortex, tiny regions of the brain dedicated to, say, inserting subordinate clauses, or turning statements into questions, or making a verb into the past tense. Grammar genes![37]

But this interpretation was soon ruled out in a groundbreaking paper of the mid-1990s by a team of London researchers who had been studying the family for a decade and who revealed that the afflicted family members' linguistic difficulties were intimately tied to a mysterious "palsy" or paralysis that limited the movements of their tongue and lips.[38] In 1998, the team published brain scans that revealed the source of this palsy: all the afflicted family members had an abnormal, underdeveloped area of the brain that controls fine motor movements: the basal ganglia, that ancient part of the brain that, as we saw earlier, plays such an important role in grooving-in the complicated, highly coordinated lip and tongue movements when a baby is learning to speak (and whose activity is shut down by oxygen starvation in high-altitude mountain climbers, and by cell death in Parkinson's disease). The family members with no voice and speech problems had normal basal ganglia; the afflicted family members had basal ganglia that was noticeably shriveled and underdeveloped.[39]

Far from suggesting a gene that controls *grammar*, the family's heritable linguistic difficulties pointed to a gene that controls *speech*—and precisely that element of speech (motor control of the articulators) that Lieberman theorized had undergone a critical change during our evolution, a genetic mutation that endowed us, and us alone, with the power to tune our voices from purely emotional cries and calls, and to shape them into articulate, spoken language through exquisitely well-planned, precisely timed movements of tongue, lips, larynx, and velum.

Strong support for *that* aspect of Lieberman's theory emerged in 2001 when the defective gene in the London family was finally isolated by a lab at Oxford University. Located on the seventh chromosome, it is a "master regulating gene" whose on- and off-switching, when an embryo is developing in the mother's womb, affects a number of "downstream" genes responsible for building parts of the lungs, heart, motor cortex, and (crucially) the basal ganglia. The Oxford team called the gene FOXP2. It turns out that we all possess two copies of FOXP2—one we get from our dad, the other from our mom. In the afflicted West London family, a transcription error in just one of those parental copies had occurred, a mix-up with two of the over seven hundred amino acid base pairs that make up that stretch of DNA. That tiny error had reduced the mobility and agility of the afflicted family members' tongues and lips enough to render speech, in the most severely affected members, almost unintelligible. (That it also affected their *thinking* was clear: all of the family members with shriveled basal ganglia had dramatically lower verbal and nonverbal IQs than the unaffected members.)[40]

Further research revealed that the FOXP2 gene is actually found in *all* mammalian species, including mice and dogs, cats and whales, chimps and orangutans, where it controls the fine motor movements involved in actions like walking and running, chewing and swallowing. As such, it has a remarkable evolutionary story to tell—a story anticipated thirty years earlier by Lieberman, and one that offers astonishing insight into the emergence of our uniquely human voice, its adaptation for speech, and our subsequent rise to the top of the food chain.

Shortly after the Oxford lab isolated FOXP2, it shared the discovery with Svante Pääbo, the head of Molecular Genetics at the Max Planck Institute for Evolutionary Anthropology in Leipzig, and a scientist famous for his research into the genes that make us human. (It was Pääbo's team that had sequenced the Neanderthal genome and revealed our familial relationship with this extinct species.) Pääbo enlisted one of his top researchers, Wolfgang Enard, to investigate FOXP2's evolutionary backstory. When Enard compared FOXP2 in mice, orangutans, gorillas, chimps, and us, he discovered that the gene had changed very little over the roughly 130 million years that separated the evolution of mice and the appearance of apes: just one amino acid change in over 100 million years. But sometime after our line branched off from the common ancestor we share with chimps (around six million years ago), the FOXP2 gene underwent *two* more mutations: two amino acid substitutions in just six million years, a major acceleration in change. The widespread appearance of this doubly mutated FOXP2 in human populations around the globe "strongly suggest[s]," Enard wrote, "that this gene has been the target of selection during recent human evolution."[41]

By "target of selection," he meant that the bodily and behavioral changes conferred by those two amino acid substitutions gave a significant survival advantage to the hominin line—*our* line. The West London family with the sluggish tongue and lips makes clear what that advantage was: a turbocharged basal ganglia that allow for high-speed, exquisitely coordinated, carefully sequenced movements of the vocal organs that only we are capable of, and that make speech possible. Enard determined that this human form of FOXP2 became "fixed" in our genome about 200,000 years ago, which is when modern *Homo sapiens* first appeared on the scene. This did not, however, rule out that our close relative, the Neanderthal, might not also have borne the mutated FOXP2. Pääbo's team analyzed Neanderthal DNA and discovered that their FOXP2 had indeed undergone the same two mutations as ours—the strongest evidence yet that Neanderthals could talk (even if their vowels were blurry). Tellingly, when the human/Neanderthal FOXP2 was "knocked-into" fetal mice by Pääbo's team, the neural pathways to the basal ganglia were enhanced and the mice, at birth, produced ultrasonic cries for their mothers that differed in pitch and "syllable" duration from those of untreated mice. [42]

But perhaps the most extraordinary finding to emerge about FOXP2 is that birds, too, possess the gene and that it expresses itself in a part of the avian brain analogous to our basal ganglia, and which controls the high-speed movements of the syrinx, tongue, and beak involved in mastering a bird's species-specific mating songs.[43] This genetic similarity between bird and human brains makes perfect sense, given that (as Darwin noted) we are among the only animal species that learn our vocalizations by exposure to adults' voices. But many scientists nevertheless greeted news of avian FOXP2 with jaw-dropped amazement because birds evolved well before mammals. Birds are actually flying reptiles (indeed, dinosaurs) and were presumed not to possess a gene that otherwise appears in the animal genome only much later in evolutionary history, with the emergence of mammals. This suggests one of two possibilities: that birds evolved *their* FOXP2 gene independently, by "convergent" evolution, and that it arose, like our FOXP2, as part of the selection pressure for vocal learning and communication crucial to birds' survival and reproduction. Or, in a scenario advanced by Lieberman, that bird and human FOXP2 originated far earlier in evolution than previously suspected; indeed, back to the Cambrian age, 500 million years ago, with the emergence of amphibians who possess basal ganglia.

Regardless, FOXP2 is emphatically *not* a "gene for language." It is, more accurately, the first gene ever found for the unique specializations of our human voice. In the twice-mutated form found in our species, it is among the best genetic evidence we possess to explain how a marginal species of small, physically weak, slow-running, hairless primates made their improbable climb, over just a few hundred thousand years, to the top of the food chain.

For Lieberman, the findings about FOXP2 and the basal ganglia are only the latest evidence to support his theory that language, far from being a purely mental phenomenon, is a physical act whose first stirrings can be traced back, hundreds of millions of years, to the oldest, air-breathing vertebrate (the lungfish), as voice (regardless of how fartlike). The exigencies of survival and reproduction gave rise to speech, some 200,000 years ago, when a series of random, but advantageous, genetic mutations led, in our early hominin line, to increased control over respiration, to the descent of the larynx, and to the powering of the basal ganglia for articulation—all anatomical accidents selected by nature for the advantages in survival and reproduction that they conferred, and which bred a bigger, better, language-capable brain. The *voice*, in Lieberman's conception, thus played the major role in creating language.

As he once put it: we "talked ourselves" into becoming human.[44]

Noam Chomsky, as we've noted, has long professed indifference to the question of where language came from—dismissing Darwinian natural selection, but declining to advance any plausible alternative. However, in the late 1990s, when the subject of language origins began to dominate linguistics, and other branches of science, Chomsky was no longer content to sit on the sidelines. In a 1999 interview, he let fall that he might see a role for Darwinian natural selection in language, after all.[45] In 2002, he published in *Science* a major paper on the subject. Coauthored by evolutionary biologists Marc Hauser and Tecumseh Fitch, the paper triggered an earthquake in linguistics for the startling minimalism (not to say bizarre reductionism) of Chomsky's idea of how language came about.[46]

Sticking to his view that our linguistic ability emerged, not for communication, but for *thinking*, he ignored the recent revelations about FOXP2, and continued to relegate Lieberman's research about the descended larynx and improved respiration to "secondary" issues concerned with the "peripheral" faculty of mere speech. Instead, Chomsky and his coauthors focused on the purely *cognitive* changes that endowed us, alone among animals, with the ability to *think* linguistically. They concluded that this ability emerged thanks to one mental operation (and *one* only): recursion.

Recursion refers to our ability to put one idea inside another. It is embodied in everything from simple adjectival phrases ("The red boat" puts the *idea* of redness inside the *idea* of a boat) right on up to complicated embedded clauses, as when discrete thoughts ("the man is walking down the street" and "the man is wearing a top hat") are combined in a single sentence ("The man who is wearing a top hat is walking down the street"). Thanks to recursion, you can just keep embedding ideas ("The man who is wearing a red top hat, which is slightly crumpled at the brim, is walking down the street and eating a slightly bruised but still delicious banana, while humming a tune made famous by Engelbert Humperdinck but which, according to musical historians, was actually written by . . ."). Words *themselves* use the recursive process by putting one speech sound (like the vowel "o") inside others ("d_g") to make "dog." Thus did Chomsky and his coauthors call recursion the "only uniquely human component" behind language, the mental engine that makes it possible to generate infinite meanings (or infinitely long sentences like the one about the man eating the banana) from a finite set of sounds (English's twenty-six vowels and consonants) and thus the *single* linguistic universal that distinguishes our speech from the grunts, hisses, moans, shrieks, and chirps of all other animals.

This was a far cry from the incredible complexity that Chomsky had always argued for in language, and which he had enshrined in his concept of Universal Grammar. His longtime acolytes were not only unconvinced—they were enraged. Told that they could now, seemingly, toss on the scrap heap the half century of work they had devoted to dissecting language for the common "deep structures" of Universal Grammar (a process that had so far brought to light only a tiny handful of supposed universals), they openly rebelled. Steven Pinker led the charge, coauthoring a response in *Cognition* with Ray Jackendoff, a leading Chomskyan. They argued that recursion, while important, was by no means the *only* "special" thing about languages, which are, they wrote, "full of devices like . . . quantifiers, tense and aspect markers, complementizers, and auxiliaries, which express temporal and logical relations."[47]

Meanwhile, tucked away in a paragraph halfway through the thirty-six-page cri de coeur was the single most salient piece of evidence to suggest that recursion could not be the whole story of language—since recursion was not even *universal.* According to Pinker and Jackendoff, there existed, in the deep Brazilian jungle, an isolated tribe of fewer than four hundred people, the Pirahã, who spoke a highly unusual language of which the most striking feature was its total lack of recursive grammar.

Pinker and Jackendoff's source was a remarkable 2005 paper written by Daniel Everett, a missionary-turned-linguist who had lived with, and studied, the Pirahã tribe for thirty years.[48] According to Everett, Pirahã speakers do not recursively embed ideas. Instead of saying, "I saw the dog that was at the beach get bitten by a snake," they would have to say, "I saw the dog. The dog was at the beach. A snake bit the dog"[49]—a little like the Motherese that Catherine E. Snow's experimental subjects used when speaking to their infants. Chomskyans immediately insisted that this aspect of Pirahã speech must reflect mental deficits (effectively *retardation*) from inbreeding—an explanation Everett instantly shot down by explaining that the tribe, although living in the most remote reaches of the rain forest, regularly refreshes its genome by sleeping with outsiders (mostly traders who ply the Amazon for Brazil nuts and wood), and are, on all other evidence, no less intelligent than any other humans. Moreover, Everett wrote, the tribe's inability to use recursion reflected not a *cognitive* constraint but a *cultural* one, since the tribe lived according to an extreme "immediacy-of-experience" principle so powerful that it affected every aspect of their lives. "When someone walks around a bend in the river, the Pirahã say that the person has not simply gone away but *xibipío*—'gone out of experience,'" Everett wrote. "They use the same phrase when a candle flame flickers. The light 'goes in and out of experience.'"[50]

This immediacy-of-experience principle, Everett said, explained the failure of missionaries like himself to convert the tribe. Told that Christ died two thousand years ago, the Pirahã lost all interest in Christianity; eventually, so did Everett, who in the late 1990s became an atheist and ceased trying to convert the tribe, focusing, instead, on studying their highly unusual language. The immediacy-of-experience principle explained why the Pirahã rejected outsiders' efforts to teach them forward-planning skills like farming or food storage, and instead still lived as hunter-gatherers unchanged since they first arrived in the Brazilian jungle some ten to forty thousand years ago.

The immediacy-of-experience principle also explained their lack of creation myths, numbers, and art—and it profoundly influenced their speech, "extending its tentacles deep into their core grammar," as Everett put it, to affect the feature that Chomsky claimed was universal to all language: recursion. Because the Pirahã accept as real only that which they can observe, in the here-and-now, their speech consists solely of direct assertions ("The dog was at the beach. It bit the man"). Recursively embedded clauses ("The dog *that was at the beach* bit the man") are not assertions but supporting, quantifying, or qualifying information—in short, abstractions, and thus impossible in the tribe's tongue.

To students of *voice*, the absence of recursion was by no means the only astounding thing about Pirahã. Unrelated to any other extant tongue, Pirahã is based on just eight consonant and three vowel sounds—an eleven-letter "alphabet" in comparison to our twenty-six—one of the simplest sound systems known. Pirahã makes up for its paucity of individual phonemes because it is a tonal language that uses the pitch of the voice to, in effect, multiply its small number of individual articulated sounds to create a sound alphabet extensive enough to make complex language possible. Mandarin Chinese is also tonal. It uses five distinct pitches. Thus, the Mandarin *ma* can mean five different things: spoken on a high, level pitch, it means "mother"; in a midrange pitch that then rises, "hemp"; in a low pitch with a slight dipping fall then rise, "horse"; with an abrupt fall from high to low, "scold." A fifth tone (called "neutral") is spoken on a weakly stressed syllable that takes its pitch from the preceding one, for questions. Tonal languages don't use *precise* pitches (you don't sing a syllable on an E-flat to mean one thing and the same syllable on an A-sharp to mean another, which would seriously inconvenience people with *amusia*, a disorder that makes them tone deaf and unable to carry a tune). Instead, tonal languages use *relative* pitch, that is, the contrast in pitch between syllables, or within a syllable. Thai, Vietnamese, most African languages, several South American and Amerindian languages are tonal. Indeed, most languages are. English and the European languages are the exceptions.

Pirahã operates similarly to Mandarin in that it uses a basic pattern of high and low tones, and combinations of those (high dipping to low; and low rising into high), but coupled with an extraordinarily complex array of stresses on syllables (by increasing the volume on a part of a word), but also syllable lengths (drawing a vowel out, or clipping it short), so that its speakers can, through combining all these elements, dispense with the individual phonemes altogether and sing, hum, or whistle conversations. All of this makes Pirahã so confounding that no outsider (trader or missionary) had mastered it for two hundred years—until Everett, an exceptionally gifted field linguist, and his similarly talented wife, Keren, arrived among the tribe as missionaries in the 1970s and, over the course of many years, achieved fluency.

Everett had been publishing papers on Pirahã for decades, but not until his 2005 article on recursion—and its challenge to Chomsky—had the wider world taken notice. Suddenly, CNN, the BBC, *Der Spiegel* magazine, and a slew of international newspapers were clamoring for an introduction to the tribe. Anthropologists, linguists, and evolutionary biologists were no less fascinated because, given the Pirahã's rejection of change, the tribe seemed to offer a snapshot of humans at an earlier time in our collective history—perhaps back to when their speech first emerged, tens of thousands of years ago. "That's what Dan's work suggests," Brent Berlin, a cognitive anthropologist at the University of Georgia, said. "The plausible scenarios that we can imagine are ones that would suggest that early language looks something like the kind of thing that Pirahã looks like now."[51]

To skeptics, Everett extended an invitation to visit the Pirahã, and test his assertions. First to accept was Tecumseh Fitch, a coauthor with Chomsky of the controversial 2002 article on recursion. With the exception of the seventy-seven-year-old Chomsky himself, Fitch was the ideal representative of the Chomsky "side." I became the sole journalistic eyewitness to this historic, linguistic showdown-in-the-jungle when, for an article in *The New Yorker*, Everett invited me to join him and Fitch in the Pirahã village. What played out in the six days and nights that we spent in the Amazon would not definitively settle the debate over whether the Pirahã are capable of recursive speech, but it offered—at least, for *me*—an unanticipated insight into the human voice, one that raised still deeper questions about the most fundamental principle upon which Chomsky built his theory of language—and about where language came from in the first place.

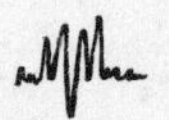

The Pirahã live in small groups of thirty to forty people spread out on the banks of tributaries that branch off the Amazon River.[52] The village

I visited with Everett and Fitch was on the Maici, a narrow, meandering tributary in the northeastern part of the country. Everett and I flew out first, in a two-seater Cessna float plane, from the former oil refinery city of Porto Velho. For an hour we soared over an ocean of unbroken rain forest, until our pilot suddenly began to descend toward the tree canopy, banking slowly until he could see a glint of silvery water between the leaves—the Maici, which proved to be a bendy river not much wider, in some parts, than the Cessna's wingspan. We landed and stepped out onto the plane's pontoon. Above us, on the steep riverbank, were some thirty or so tribe members: short, dark-skinned men, women, and children, some with babies on their hips, others clutching bows and arrows. They responded to the sight of Everett with a greeting that sounded like a profusion of exotic songbirds, a melodic chattering scarcely discernible, to the uninitiated, as human speech.

Everett answered them in the tongue's choppy staccato: "*Xaói hi gáisai xigíaihiabisaoaxái ti xabiíhai hiatíihi xigío hoíhi.*" He was telling them that I would be "staying for a short time" in the village.

The men and women answered in an echoing chorus, "*Xaói hi goó kaisigíaihí xapagáiso.*"

Everett turned to me. "They want to know what you're called in 'crooked head.'"

This was the tribe's term for any language not Pirahã, and it was a clear pejorative. The Pirahã consider all forms of human discourse other than their own to be laughably inferior, and they are unique among Amazonian peoples in remaining completely monolingual. Told my name, they playfully tossed the sound back and forth among themselves, altering it slightly with each reiteration, until it became an unrecognizable syllable. They never uttered it again, and instead gave me a lilting Pirahã name: Kaaxáoi, that of a tribe member, from a village downriver, whom they thought I resembled. "That's completely consistent with my main thesis about the tribe," Everett told me. "They reject everything from outside their world. They just don't want it, and it's been that way since the day the Brazilians first found them in this jungle in the 1700s."

Fitch, who was traveling with his cousin Bill, a sommelier based in Paris, arrived a few hours later in the Cessna, which had circled back to Porto Velho to fetch them. The tribe members surrounded the cousins as they stepped onto the riverbank. Having traveled widely together to remote parts of the world, the Fitches believed that they knew how to establish an instant rapport with indigenous peoples. They brought their cupped hands to their mouths and blew loon calls. The Pirahã looked on stone-faced. Bill made a popping sound by snapping a finger against the opposite palm. The Pirahã remained impassive. The cousins shrugged sheepishly. "Usually you can hook people really easily by doing these funny little things," Fitch said later. "But the Pirahã kids weren't buying it, and neither were their parents."

"It's not part of their culture," Everett said. "So they're not interested."

The "village" was as Everett had described it in *Current Anthropology*: seven huts made of palm fronds propped on sticks. Mud floors, no decorations, no walls. During his first two decades living with the tribe, Everett and his family had slept in a tent on the edge of the village. But in 1999, he had built a two-room, eight-by-eight-meter, bug- and snake-proof house from ironwood and equipped it with a gas stove, generator, water filtration system, and other amenities. It was here that the four of us slept and where some twenty Pirahã gathered outside each morning to volunteer as research subjects. The testing was carried out in a separate hut, propped on tall stilts, that Everett used as his office.

Fitch had brought a laptop loaded with programs based on the "Chomsky hierarchy," a system for classifying types of mental "grammar," the term Chomskyan linguists use to describe the specific mental operation that turns abstract thought into language. Fitch began with the simplest grammar, one which would determine whether the Pirahã were capable of learning a basic pattern of sounds. The pattern involved a male voice uttering one syllable (*mi* or *doh* or *ga*, for instance), followed by a female voice uttering a different syllable (*lee* or *ta* or *gee*). (Male and female voices were used in order to make a clear differentiation between the syllables, much as you would use differently colored blocks if asking a test subject to group them into specific patterns.) When the voices spoke a "correct" construction (one male syllable followed by one female), an animated monkey head at the bottom of the computer screen would float to a corner at the top; "incorrect" constructions (anytime one male syllable was followed by another male syllable or more than one female syllable) would make the monkey head float to the opposite corner. Fitch wanted to know if, in repeated trials, the Pirahã could figure out, and learn, these patterns, a mental function basic to all language comprehension. He mounted a small digital movie camera to the top of the screen so that he could film a test subject's eye movements. In the few seconds' delay before the monkey head floated to either corner, Fitch hoped that he would be able to determine, from the direction of the subjects' unconscious glances, if they were learning the "grammar." The experiment, using different stimuli, had been conducted, at Harvard, with undergraduates and monkeys, all of whom passed the test. "My expectation is that they're going to act just like my Harvard undergrads," Fitch said. "They're going to do exactly what every other human has done and they're going to get this basic pattern. The Pirahã are humans—humans can do this."

But in the first few days, each tribe member, seated in front of the laptop, would watch the monkey head and not respond to the audio cues. Everett, who acted as translator during the experiments, explained to Fitch that the barrier was cultural, not cognitive: all of this was so alien to the Pirahã experience, they simply didn't understand what was expected of them. Their eyes seemed to go everywhere. Fitch asked Everett to tell the subjects to "point to where they think the monkey is going to go." Everett explained another cultural constraint: the Pirahã don't point. Nor do they have words for right and left. They tell others to head "upriver" or "downriver," or "to the forest" or "away from the forest." On Fitch's request, Everett told a male subject to say whether the monkey head was going upriver or down. "Monkeys go to the jungle," the man responded, quite logically. Other subjects expressed a natural curiosity about the floating head: "Is that rubber?" "Does this monkey have a spouse?" "Is it a man?" When Fitch's computer froze in the jungle humidity, he stormed off to the main house to fix the problem. "This is typical of fieldwork in the Amazon, which is why most people don't do it," Everett told me. "But the problem here is not cognitive; it's cultural," he reiterated. He gestured toward the Pirahã man at the table. "Just because we're sitting in the same room doesn't mean we're sitting in the same century."

On the fourth day, a girl of perhaps sixteen—focused, alert, and calm—seemed to grasp the grammar, her eyes moving to the correct corner of the screen in advance of the monkey's head. Fitch, excited, tested her on a higher level of the Chomsky hierarchy, a "phrase-structure grammar," in which correct constructions consist of any number of male syllables followed by an equal number of female syllables. In their 2002 paper on recursion, Chomsky, Hauser, and Fitch had stated that this grammar, which makes greater demands on memory and pattern recognition, represents the minimum foundation necessary for human language. Fitch watched closely, trying to discern whether the girl's eye movements indicated that she was learning the pattern's rules. It was impossible to say with the naked eye. He would have to take the footage back to his lab at the University of Edinburgh, and have the film vetted by an impartial postdoc, who would "score" the images on a timeline synchronized to the soundtrack of the spoken syllables. Only then would Fitch be able to say with certainty whether the subject's eyes had anticipated the monkey head—or merely *followed* it.[53]

In our remaining two days in the village, Fitch failed to find another subject as promising as the girl, but he pronounced himself satisfied with the data he had collected. He was even willing to concede that culture had perhaps constrained certain aspects of the tribe's speech. "But as far as the Pirahã disproving Universal Grammar?" he said. "I don't think anything I could have seen would have convinced me that that was ever anything other than just the wrong way to frame the problem." Everett insisted that it was the *only* way to frame the problem. He added that the endless search for Chomsky's "universals" had destroyed the beauty of linguistics, by ignoring the things that make languages so fascinatingly *different* from one another—the quality that had captured the imagination of the world's first anthropological linguist, Edward Sapir, when he studied the indigenous Indian tribes of the American West at the dawn of the twentieth century.

"When I went back and read the stuff Sapir wrote in the twenties," Everett said, "I just realized, hey, this really is a tradition that we lost. People believe they've actually studied a language when they have given it a Chomskyan formalism. And you may have given us absolutely no insight whatsoever into that language as a separate language."

I received an entirely different view of Pirahã, from that of either Fitch or Everett, on my final night in Brazil, after the Cessna flew us back to Porto Velho. There, I met in the lobby of our hotel with the only other outsider who could speak the language: Everett's estranged wife, Keren. She and Everett had separated in 1999, after twenty years of marriage, in part because Keren kept her Christian faith and continued trying to convert the tribe. An elfin woman in her fifties, Keren still lived with the Pirahã six months of the year, but had chosen to withdraw to Porto Velho during our time in the jungle. Though her primary interest in the tribe remained missionary, Keren was also fascinated by their language. Like Everett, she had studied formal linguistics and was well acquainted with the Chomsky program. But her insights into Pirahã transcended the bitter debate over recursion and Universal Grammar, tree structures, X-bars, and phrase structures (Chomsky's recondite terms)—for the simple reason that she did not think that these were important to understanding, or mastering, the language. The key, she had recently come to believe, was in the tribe's *singing*: the way that they can drop consonants and vowels altogether and communicate purely by variations in pitch, stress, and rhythm—the prosody.

"This language uses prosody much more than any other language I know of," Keren told me. "It's not the kind of thing that you can write, and capture, and go back to; you have to watch, and you have to feel it. It's like someone singing a song. You want to watch and listen and try to sing along with them. So I started doing that, and I began noticing things that I never transcribed, things I never picked up when I listened to a tape of them, and part of it was the performance. So, at that point I said, 'Put the tape recorders and notebooks away, focus on the person, watch them.' They give a lot of things using prosody that you never would have found otherwise. This has never been documented in any language I know." Aspects of Pirahã that had long confounded Keren became clear, she said, through the music of their voices. "I realized, Oh! That's what the subject-verb looks like, that's what the pieces of the clause and the time phrase and the object and the other phrases feel like. That was the beginning of a breakthrough for me."

Keren's emphasis on the music of Pirahã brought to mind Darwin's insight that language evolved from song, as strings of expressive tones arranged in melodic contours, into which words were gradually introduced. I wondered if Everett and Fitch, and the scores of linguists who had since weighed in on Everett's *Current Anthropology* paper were, by bickering endlessly over recursion and syntax, missing something far more important that the Pirahã had to tell us about ourselves and where we might have come from.

Singing, Keren went on, had been the breaking point for Everett. His frustration at having to "start all over again" with the language led to his leaving the Amazon and their marriage and taking a full-time teaching job in Manchester, England. "Pirahã has just always defied every linguist that's gone out there," she said, "because you can't start at the segment level and go on. You're not going to find out anything, because they really can communicate without the syllables."

As Keren spoke, I was reminded of one of my last evenings in the Pirahã village when a strange sound began to waft into Everett's house: a voice singing a clutch of notes on a rising, then falling scale. Everett and Fitch, arguing over an abstruse element of Chomskyan linguistics, ignored the voice. I stepped outside and made my way through the gloaming toward the sound. It was coming from a hut on the edge of the village. I crept closer and saw that it was a woman, winding raw cotton onto a spool. For twenty minutes, she intoned this extraordinary series of notes, over and over, her voice like a muted horn. A toddler played at her feet.

When I returned to the house, I asked Everett about what I had seen and heard. He said something vague about how tribe members "sing their dreams." But when I now described the scene to Keren, in the Porto Velho hotel, she grew animated and explained that this is how the Pirahã teach their children to speak. The toddler was absorbing a lesson in the language's all-important prosody, through the mother's endless repetition of that melody—a living example, one might argue, of Sapir's view of language, not as an inborn instinct, but as a *cultural* skill passed, from generation to generation, through the medium of the voice. Only recently—that is, since researching this book and learning of how the mother's voice begins tutoring a child in language even before the baby leaves her womb—have I come to see that Pirahã woman and her toddler in a still wider context, one that highlights the underappreciated role—the *incalculable* role—that women's voices played in our evolutionary ascent, as *the* primary acoustic signal that prepares the developing human brain for language.

Which is not to suggest that the *male* voice has been entirely inconsequential in our evolution, given the role it played, and continues to play, in the sexual signaling critical to the propagation of our species.

FIVE

SEX AND GENDER

In October 1994, NBC-TV aired an episode of *Seinfeld* that began with Jerry expressing amazement to Elaine that one of their friends, Noreen, was hitting on him over the phone. "No*reen*?" Elaine says. "But she's got a new *boyfriend*!" Mystified, Elaine calls Noreen to feel out the situation. She is several seconds into the call (we hear a distinctly female voice coming through Elaine's earpiece), when she asks, "Noreen, were you hitting on him?" The camera cuts to the person on the other end of the line—a short, middle-aged, bald man: Dan, Noreen's boyfriend, whom Jerry and his friends have dubbed a "high talker." Elaine, fooled by his voice, has given away Noreen's secret. "You're saying that Noreen was hitting on *Jerry Seinfeld*?" Dan says, aghast.[1]

In real life, such mistakes are rare. That's because the human voice is unique, in the animal kingdom, not only in its specialization for speech, but for its sexual dimorphism—the way it splits along gender lines. All other mammals are vocally monomorphic: their roars, barks, meows, and baahs sound the same whether made by a male or a female of the species. Even our closest primate kin, chimpanzees and bonobos, display less sexual dimorphism of voice than we do—only a few semitones difference. Human males speak at a pitch that is, on average, a full *octave* below women, twelve semitones, a big difference.[2]

You can, of course, find outliers. Writer David Sedaris has complained that his unusually high voice invariably makes room service operators politely call him "ma'am."[3] Some deep-voiced women are frequently mistaken for the opposite sex.[4] But these are exceptions that prove the rule (which is why they strike us as anomalous—and as fodder for comedy). In one study of over six hundred male and female undergraduates, researchers found zero overlap in vocal pitch between the men and women: the most baritone of the females spoke at a higher pitch than the highest-voiced male.[5] When it comes to the human voice, nature wanted to establish a clear division between the sexes.

The human voice actually starts off, in childhood, as sexually monomorphic.[6] But the male voice deepens at puberty when, around age thirteen, the testicles emit a time-released blast of hormones, known as androgens (including testosterone). These chemicals cycle through the blood and bind to androgen receptors on muscles and tissues throughout the body, changing their gene expression and causing them to explode in growth. Girls also produce androgens, just less of them (because they don't have testes). Heavy androgen exposure is why teen boys, at puberty, develop larger muscles than girls and (on average) grow taller. The larynx happens to be rich in androgen receptors so the cartilages, ligaments, and muscles all become larger at puberty (the thyroid cartilage making its presence known as a pointed protuberance on the outside of the male neck: the Adam's apple) and the vocal cords triple in size: lengthening, widening, and thickening, slowing their air-chopping rate, lowering the voice's perceived pitch.

Not surprisingly, it takes adolescent boys a while to adapt themselves to the unfamiliar physical dimensions of the "new" instrument inside their neck: until they groove a revised set of neural instructions into the basal ganglia, they're prone to emitting sudden high-pitched squeaks and squeals. At the same time, the widening and lengthening of the neck, and expansion of the mouth and nasal cavities, change the resonances of the vocal tract, boosting lower overtones, muting higher ones, thus adding a "dark" or "rich" timbre to many adult male voices. Finally, the muscles of the diaphragm and intercostals also increase in size and strength under the androgen surge, lending to male voices' greater volume, on average, than that of females. So, the dimorphism of the human voice is not purely a matter of pitch; there are also distinct differences in timbre and volume between men and women. But pitch tends to be the one we notice most. Importantly, girls' voices also get lower at puberty (because of androgen secretion from the adrenal glands), just not as dramatically. The female vocal cords increase only slightly in size—about 32 percent to men's 68 percent—so women's voices descend only a semitone or two, and thus retain much of the high, clear sound they had before puberty, and which preadolescent boys share.

That high, clear sound was so prized by papal church choirs of late sixteenth-century Italy that it led to a barbaric practice: the removal of young boys' testicles so that their voices would not change at puberty. The church could have avoided this drastic expedient simply by using girls and women as singers—except that females were forbidden, by canon law, from performing in churches. The boys were usually operated on between the ages of seven and nine. As adults, these singers, known as castrati, had vocal cords that remained small and thin, but their bodies tended to grow close to adult male size.[7] This resulted in an unusual hybrid sound: a female pitch, but a male timbre and power. One contemporary writer described the castrati as sounding "as clear and penetrating as that of choirboys but a great deal louder with something dry and sour about it yet brilliant, light, full of impact."[8] A craze for the voice occurred in the mid-seventeenth century with the rise of Italian opera, for which castrati parts were expressly written. At the craze's peak, an estimated four thousand boys a year were castrated—many of them poor children whose parents gave them over to gelding in the hope that they would become successful singers and lift the family from poverty.[9] In 1861, the practice was finally stopped, but castrati continued to be produced, for the private delectation of the pope, in his Sistine Chapel. The last Vatican castrato, Alessandro Moreschi, officially retired in 1913.

While we know *how* the human voice becomes sexually dimorphic at puberty, we don't know exactly *why*. But given that the differences emerge at sexual maturity, the safe assumption is that it has to do with mating and reproduction.

In *The Origin of Species*, Darwin said that the successful wooing and winning of reproductive partners involves two quite separate mechanisms: attracting mates through the display of seductive ornaments (like the male peacock's tail or the finch's fancy songs), but also driving off, or vanquishing, same-sex rivals. The latter mechanism, which later evolutionary biologists called "contest competition," almost always involves males battling other males (surprise!) for the favors of females, and this led to the evolution, on male bodies, of what Darwin called "special weapons,"[10] including the antlers on stags (who run at each other and butt heads at mating time), leg spurs on roosters (who slash at other males to neutralize them as sexual competitors), and long incisor teeth on male chimpanzees (who flash these fangs at other males when competing over a female). Some evolutionary biologists explain the deepened male voice in *Homo sapiens* as one of these special weapons. While the voice can't physically maim or kill, like an antler, leg spur, or fang, it is weaponized in aggressive encounters by a dropped pitch, which creates the illusion of greater body size. Thus, those hominin males who, through a quirk of their genetics, happened to possess larger, thicker vocal cords than the norm, and thus naturally deeper voices, would have had an inborn advantage over males with slightly smaller vocal cords and thus higher voices. Scaring off their squeakier sexual rivals, deep-voiced males won the contest competition for mates, and thus the genes for deep-voiced males were sexually selected in our ancestors. The male voice went south.

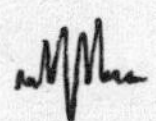

David Puts, a professor of anthropology and evolutionary psychology at Penn State University, has spent two decades exploring the sexual dimorphism of the human voice and has concluded that, in terms of vocal pitch, twenty-first-century males differ little from our ape ancestors. In a test of nearly two hundred male college students, Puts showed that men adopted a lower pitch when speaking to males whom they perceived to be inferior to them socially and physically. In turn, those "inferior" men (who actually rated *themselves* as less dominant than their deeper-voiced peers) raised their vocal pitch, deferentially, when speaking to guys they saw as higher on the social and physical hierarchy.[11] In the same study, the dominant, lower-voiced men reported a greater number of sexual encounters in the previous year than the men with slightly higher voices. Like our hominid ancestors, then, baritone-voiced males, it seems, get more sex—but not only because they "drive off," or vanquish, their higher-pitched male rivals in a modern-day "contest competition" for mates. It's also a matter of female taste.

In controlled studies, women consistently rate men with lower voices as more sexually attractive than men with higher voices.[12] Puts explains this by invoking the other Darwinian mechanism by which animals choose sexual partners: "mate selection," the notion that certain secondary sexual characteristics—like a tall, muscular, symmetrical body (suggestive of strength and health)—indicate the superior genes you'd want to pass to your offspring. Indeed, research shows that deep male voices correlate with high testosterone levels (makes sense: the more androgens, the bigger and thicker the vocal cords)—and testosterone strengthens immune response. Thus, females, in selecting deeper-voiced guys, are simply obeying an imperative wired into their brains by evolution: they choose the reproductive partner most likely to give their children the genetic advantage of greater disease resistance. But a fascinating complication emerged from Puts's research—and one that should cheer up the high-talking Dans of this world.

Female college students have a different reaction to male vocal pitch depending on where the women are in their menstrual cycle.[13] When ovulating, and thus at greatest risk for getting pregnant (and at their most libidinous), they prefer low-pitched male voices; during the least fertile part of their cycle (that is, right after their period, when they're feeling less erotically minded), the same women prefer men with voices a few semitones higher. The deeper-voiced males were, it turned out, attractive as short-term hookups, or imaginary partners (during ovulation, the women reported fantasizing more freely about sex outside their monogamous relationships), whereas the higher-voiced guys were attractive as *actual* long-term mates. Apparently, women have evolved an understanding that men with deeper voices also display an array of less desirable traits typical of excessive masculinization through overexposure to testosterone: specifically, the kind of high sex drive likely to make them seek other sexual partners.[14] Hence the female preference (at least, when they're not ovulating), for men whose slightly higher voice indicates a more settled, more *monogamous* temperament. A guy, in short, who will stick around to help raise their offspring.

These findings are consistent with research that shows women prefer men who are less overtly masculine in *all* ways, not just voice. (In tests, women consistently rate as most attractive leanly muscled men over bulked-up body builders; men who are tall but not too tall; men with smooth-shaven faces over bushy beards; and men with a facial bone structure that lies more on the feminine end of the spectrum—Johnny Depp, say, over Josh Brolin).[15] According to Richard O. Prum, a leading evolutionary biologist at Yale University, female mate choice is a primary force that shapes animal species, including our own: while male-on-male contest competition tends to weaponize aspects of the male body and temperament, mate selection (as determined by female choice) tends to *de*-weaponize it, for the understandable reason that the male special weapons (fangs, leg spurs, antlers, and deep, intimidating voices) that evolved for contest competition can also be used as tools for sexual coercion of females. By choosing reproductive partners who are less domineering (and dangerous), females produce male offspring with less coercive weaponry.[16] Hence the shrinking, in *Homo sapiens*, of the massive incisor fangs that male chimps use both to drive off reproductive rivals and to force sex on females in acts that, to us, look distressingly like rape. This is likely how our species has ended up with male voices significantly lower than that of females, but nowhere near the rumbling growl of gorillas. With their evolved attraction to voices that are low (but not *too low*), women have dialed-up the average pitch of the male voice from that of our primate ancestors, even at the cost of a slightly weaker immune system in their offspring.

Darwin cited gibbons as a living example of how certain ape species use vocal melodies, in a complex duet, for courtship. If gibbons are any guide, the melodic elasticity of the voice is a big romantic draw—especially its ability to soar into the upper registers, in clear high notes that (in males) indicate that the vocalizer is not only a fierce warrior, but also a sensitive, nurturing, romantic soul.

The castrati craze of the seventeenth century suggests that a high, yet still recognizably *male*, voice has a certain appeal to the human aesthetic sense—and perhaps especially to females in their childbearing years. Here, I would point to the enduring trend in modern pop music for extremely high male singing voices. The prepubertal Justin Bieber offers a current example, but a high, bell-pure vocal tone was also a major feature of Paul McCartney's singing voice during the Beatlemania years, a sound that drove female audiences into paroxysms of "answering" screams. Elvis Presley, a few years earlier, provoked an identical reaction in girls with his similarly high, clear, grit-free voice on songs like "That's All Right Mama." Singers of a still earlier generation, Bing Crosby and Frank Sinatra (male pop singers who, with the advent of radio in the 1930s, were the first to trigger mass hysteria in teen girls), were not especially known for hitting stratospheric high notes, but their crooner-style did take down the voice's volume, stressing the nuzzling *n* and *m* sounds, and thus making it a signal not of threat or dominance, but of intimacy, a lilting, aural caress, a thing of love, not war.

The evolutionary psychologist Geoffrey Miller draws an explicit connection between male music making and sexual success, pointing out that the singer-guitarist Jimi Hendrix had "sexual liaisons with hundreds of groupies, maintained parallel long-term relationships with at least two women, and fathered at least three children in the U.S., Germany, and Sweden." Hendrix was, Miller says, a living example of Darwin's theory that "sexual ornaments" (in this case the ability to make gorgeous music), greatly enhance an organism's ability to spread its DNA into subsequent generations: "Hendrix's genes for musical talent probably doubled their frequency in a single generation, through the power of attracting opposite-sex admirers."[17] Daniel Levitin, in *This Is Your Brain on Music*, makes the same argument, using Led Zeppelin singer Robert Plant as his example. Levitin quotes Plant speaking about the band's famously (or infamously) hedonistic concert tours of the 1970s. "Whatever road I took," Plant once said, "the car was heading for one of the greatest sexual encounters I've ever had."[18] The band's appetite for groupies is, indeed, legendary. But it is worth noting here that Plant's voice, in its glass-shatteringly high pitch and ear-crushing lung power, seems to have possessed many of the qualities that eighteenth-century critics praised in the castrati. So, ironically, Plant's sexual success might have derived from the message his singing voice (like Bieber's and McCartney's) sent to female fans of a *lack* of sexual threat, or at the very least the promise of monogamous devotion—a dishonest signal, if ever there was one. Jimi Hendrix, meanwhile, sang in a far lower pitch, and darker timbre, than Plant or, indeed, most male rock stars. But with his Stratocaster, he made abundant use of the high notes: he let his guitar solos, with which he mimicked the sound of a keening human voice, do the wooing.

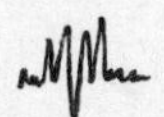

Unlike in most bird species, where males do all the sexual singing, vocal wooing in humans is a two-way street, and the power of women's voices to charm and seduce has been recognized since (at least) ancient Greece. Homer's hero, Odysseus, avoided the fatal lure of the Sirens' song only by lashing himself to his ship's mast, his crew stopping up their ears with beeswax. (Odysseus refused earplugs because he was determined to hear what no other man had heard—and survived: the sound of the Sirens' voices.) Homer did not describe what the Sirens actually sounded like, beyond vague allusions to their "celestial" tones, but if modern science is any guide, their voices were somewhat high-pitched and slightly breathy.

That, at least, is what laboratory tests tell us about the voices that college-aged males consistently rate as the most sexually attractive in females.[19] Evolutionary biologists believe that men developed a preference for a relatively higher female pitch because it carries a message of youth and thus reproductive health (women's voices deepen with age and the onset of menopause).[20] Breathiness, meanwhile, is a positive sexual signal because it derives from a tiny anatomical difference between the male and female larynx that emerges at puberty: women's vocal cords do not close, completely, at the back of the larynx, when phonating. The minuscule gap between the membranes allows a small amount of air to escape "unchopped," lending a whispery edge to women's speech.[21] Thus men, in their search for the most feminized (and thus most fertile) partner, are subconsciously attracted to women with a higher pitch and some telltale purring turbulence in their vowels and voiced consonants. This would help to explain the popularity of the actress Marilyn Monroe, who deliberately exaggerated all aspects of human sexual dimorphism, with her spiked high heels, curvy walk, hydraulic brassieres, cinched-in waist, heavy makeup—and high-pitched, whispery, kewpie-doll voice; a voice that, in sexual-selection terms, advertised "youthful" fertility to the point of parody.

As Monroe's vocal exaggerations suggest, the sexual dimorphism of the human voice is not purely a result of anatomy. Social expectations, cultural pressures, gender norms, and individual psychology all affect how men and women tune their vocal pitch and voice quality. A 1995 study by a Dutch research team showed that Japanese women speak several semitones higher than women in Western countries, and at a softer volume: a voice consistent with the traditionally approved role of women in Japanese society, where a code of female conduct, called *onnarashii*, rewards "modesty, innocence, dependence, subservience."[22] Japanese men reinforce this socially imposed divide by speaking at a pitch a few semitones lower than men in the West. North American and European men also drop their voices into an artificially low pitch to advertise their maleness, but usually when in same-sex groups, like sports bars and locker rooms, where they can be heard booming at each other in passive-aggressive displays of "camaraderie." This super-masculinized voice is especially off-putting when speakers transport it from all-male enclaves and use it in their personal or professional lives.

Experimental psychologist Nalina Ambady showed just how off-putting. She used a device called a low-pass filter on recordings of surgeons speaking to patients in routine office visits. The filter removes all language content from speech, but preserves tone and pitch—the prosody. Listener-judges were asked to rate the voices for "warmth," "anxiety," "concern," "interest," "hostility," and "dominance." Using only those ratings—and with no knowledge of the doctors' professional histories or competence—Ambady was able to predict, with 100 percent accuracy, which surgeons had been sued and which had not.[23] A *dominant* tone ("deep, loud, moderately fast, unaccented, and clearly articulated") was the giveaway for who got sued. The listener-judges reported that these voices conveyed "lack of empathy and understanding." Surgeons with "warm," "sympathetic," or "concerned" voices faced no malpractice claims—even when they had histories of harming patients through incompetence or negligence. The off-putting voices that triggered vengeful lawsuits were defined largely by their excessive baritone pitch.

Women have hardly been deaf to the message of power that a lower voice bestows on the speaker and they have, for some time now, been adjusting their voices downward. Speech pathologists in Australia recently compared the voices of young adult women recorded in 1945 with the voices of Australian women recorded, under identical conditions, in 1993.[24] The team discovered that in the intervening fifty years women's pitch had dropped, on average, almost two semitones—a significant change. Studies of American, Canadian, and Swedish women showed that they speak in the same lowered range as Australian women. Furthermore, earlier studies (from the 1920s and 1930s) reveal that women once spoke at a still higher pitch than that of women in the 1940s—as high as 318 cycles per second, more than half an octave higher than women today.

The steady drop in female vocal pitch parallels, and was almost certainly driven by, the seismic changes in women's social status over the last century, beginning with the international suffrage movement, which by 1920 had earned women the right to vote in countries throughout the West. Female vocal pitch continued to drop during the Second World War, when women were liberated from the roles of housewife and mother and called on to fill positions of authority in the workplace vacated by men who had gone to fight. Women were also recruited as reporters and announcers on radio, where they were obliged to speak in a lower pitch than normal because of audio equipment designed for broadcasting men's voices (speech in higher registers tended to distort).[25] C. E. Linke, a voice specialist at Tulane University, says that the "psychosocial influence" of popular women radio personalities during the war helped to drive down the fundamental frequency of female voices across society.[26]

Hollywood also played a critical role. Movies of the mid-1940s (reflecting women's wartime liberation from the home) were suddenly filled with fast-talking, wisecracking dames[27]—female characters who, in the pitch, pace, and content of their speech, could hold their own in verbal jousts with their male costars (Rosalind Russell in *His Girl Friday*, Katharine Hepburn in *Bringing Up Baby*). Lauren Bacall was a nineteen-year-old former model who made her screen debut in the 1944 film *To Have and Have Not* and instantly became famous for speaking in deep, bassoon-like tones never before heard in a leading lady. In her 1978 memoir, *By Myself*, Bacall revealed that this was not her natural pitch: she trained her voice for the role by reading aloud in a low, loud voice, for hours on end.[28] She did this at the urging of the film's director, Howard Hawks, who hoped to create with Bacall's character a vision of the newly liberated woman, one whose "masculine" vocal pitch and assertive attitude only enhanced her erotic appeal—"a girl," as Hawks later said, "who is insolent, as insolent as Bogart, who insults people, who grins when she does it, and people like it."[29]

People did like it. A huge star of the 1940s, Bacall became a leading role model for feminists of the 1970s, including Germaine Greer, who extolled the empowering message conveyed by the actress's deep voice and badass bearing. But Greer also lamented what happened to Bacall's career when the war ended and movies began to reflect a society in which, as Greer put it, "women were back in the bedroom and the kitchen, working on the baby boom." Bacall's career foundered in the 1950s as movie directors began casting actresses with voices whose pitch and timbre communicated something very different to Bacall's "insolent" and sexually assertive characters of the 1940s. Even Hawks capitulated to the new postwar zeitgeist: his "next starring ladies," Greer wrote, "would be Marilyn Monroe and Joan Collins in the cinch-waisted, pointy-breasted, simpering 1950s."[30] According to Gloria Steinem, another leading feminist, these "simpering" actresses set a dangerous example for women seeking equal pay and equal rights with men. In a 1981 essay in *Ms.* magazine, she warned: "A childlike or 'feminine' vocal style becomes a drawback when women try for any adult or powerful role."[31]

There is abundant evidence that the female voice, no matter how it's pitched, is a drawback in Western societies. The classical scholar and feminist Mary Beard has traced how women's voices have been systematically silenced since antiquity. She begins her book *Women & Power* (2017) with a scene from Homer's *Odyssey*, when Penelope, wife to the absent wanderer Odysseus, asks a ballad singer performing for a large group of her suitors to play a less lugubrious number. Her son, Telemachus, barks at her: "Mother, go back up into your quarters, and take up your own work, the loom and the distaff . . . speech will be the business of men."[32] Telemachus uses the term *muthos* for "speech," which in this context specifically refers to *authoritative public speech*, "not," Beard adds sardonically, "the kind of chatting, prattling or gossip" that women were believed to be solely capable of. Ancient attitudes to consigning female voices to the domestic sphere have changed less than you might expect (especially if you're male). Today, when women have made considerable strides in the corporate, professional, and political worlds, their voices continue to be shut down, or blandly appropriated by men. Beard reprints a cartoon from the satirical magazine *Punch* in which a CEO addresses five corporate underlings, only one of whom is a woman. "That's an excellent suggestion, Miss Triggs," the CEO says. "Perhaps one of the men here would like to make it." The cartoon is thirty years old, but many women I know, young and old, find it a little too accurate to be funny.

When men are not blithely ventriloquizing ideas stated moments earlier by the women in their midst, they are speaking over them. In 2017, researchers at the Pritzker School of Law at Northwestern University studied transcripts of U.S. Supreme Court oral arguments going back fifteen years. Male justices and male lawyers "interrupt the female justices approximately three times as often as they interrupt each other."[33] Recently, the male tendency for oblivious, blowhard-like interruption and holding-forth has been called "mansplaining," a term inspired by Rebecca Solnit's viral essay, *Men Explain Things to Me* (2014), which begins with her tale of being lectured to by a wealthy alpha male about a book on

Eadweard Muybridge.[34] Solnit is too nonplussed to stop the man and explain that the book he is talking about was written by *her.* Solnit's friend, Sallie, tries to interject: "That's her book." Sallie has to say it three or four times before the man finally hears—and falls silent. Solnit admits that "people of both genders" can hold forth self-importantly, "but," she adds, "the out-and-out confrontational confidence of the totally ignorant is, in my experience, gendered."[35]

No contemporary discussion of women's voices can fail to discuss vocal fry. Also known as "creaky" voice, the fry is a low-pitched, crackling, croaking sound—a little like frying bacon, from which the voice got its name. Scientists first took note of its ubiquity in young women in a fall 2010 paper in *American Speech.*[36] A raft of news reports followed describing (and decrying) the phenomenon. Mark Liberman, a professor and creator of a popular linguistics website, *Language Log*, meticulously catalogued references to vocal fry in the lay press, including an announcement, by *This American Life* host Ira Glass, about the flood of hate mail that NPR was receiving about Glass's young female cohosts.[37] "Listeners complain about their 'vocal fry,'" Glass said, and went on:

> These are some of the angriest emails we ever get. They call these women's voices unbearable, excruciating, annoyingly adolescent, beyond annoying, difficult to pay attention to, so severe as to cause discomfort.

The reaction wasn't confined to NPR listeners. Scholarly articles proliferated, measuring the sociological ill effects of vocal fry on speakers. The mannerism made them sound "hesitant, nonaggressive, and informal."[38] Owing to how irritating it is, the fry "may undermine the success of young women."[39] Nevertheless, vocal fry has continued to spread, like a contagion.

That vocal styles *are* highly contagious was a fact I first learned about years ago when reading Tom Wolfe's *The Right Stuff*, which documents the training of the first NASA astronauts, who were all recruited from the ranks of test pilots for the Army Air Force, including test pilot Chuck Yeager, who had become famous in 1947 as the first human to break the sound barrier in a supersonic jet. Yeager, who was born in a coal-mining community in West Virginia, spoke in an unhurried drawl so effective in communicating imperturbability from the cockpit that his awestruck fellow pilots couldn't resist imitating it. "Military pilots and then, soon, airline pilots, pilots from Maine and Massachusetts and the Dakotas and Oregon and everywhere else, began to talk in that poker-hollow West Virginia drawl," Wolfe wrote, "or as close to it as they could bend their native accents. It was the drawl of the most righteous of all the possessors of the right stuff: Chuck Yeager."[40] It's also the drawl that every (male) commercial pilot in North America speaks in to this day, regardless of where they were born—a voice that instantly communicates to all on board that the guy behind the controls has *got this*. Tom Farrier, the former director of safety for the Air Transport Association, recently posted reflections on the Yeager-drawl, including the observation that most pilots maintain it even during the most dire emergencies.[41] This was recently demonstrated by Captain Chesley (Sully) Sullenberger when double engine failure on takeoff from LaGuardia Airport obliged him to land his aircraft in the Hudson River. Cockpit recordings (available on YouTube) reveal that throughout the four-minute ordeal Sullenberger's voice never loses its laconic, low-pitched, slightly drawling tone.[42] Pure Yeager.

Investigators on a similar epidemiological quest to learn the origins of today's vocal fry traced *its* viral popularity to reality television star Kim Kardashian, whose immensely popular show, *Keeping Up with the*

Kardashians, debuted in 2007. The first reports of the vocal fry epidemic began to appear as the show reached peak viewership in 2010[43]—a synchrony of Kardashian popularity and creaking-voice-ubiquity that certainly suggests Kim (who speaks in a perpetual, pronounced vocal fry) is the epidemic's Patient Zero. But nobody has been able to explain why, exactly, her speaking style proved so catching—beyond the general tendency for people to emulate the behavior of celebrities and to turn those behaviors into short-lived fads. This is no short-lived fad. Kim's vocal fry has operated more like Yeager's drawl—it has colored the way women speak across the culture, which suggests that it has tapped deeply into how women want to be perceived. In short, it's time to start thinking of it as a zeitgeist-defining phenomenon. But what larger truth could vocal fry be telling us about women today, how they feel and what they want to project?

My initial explanation focused on how the fry reduces the voice to a monotone crackle, erasing prosody, and thus emotion, from speech. This seemed to be how Kim used it. Working in concert with her heavily made-up, masklike facial features, her monotonic voice projected an impression of blasé detachment, which contributed to her "brand" as a Beverly Hills diva whose wealth cocooned her from ordinary earthly concerns (and thus emotions). She had attained the condition to which the popular culture urges everyone to aspire—indeed, to *keep up with*: unimaginable wealth and fame with zero effort and no discernible talent. My theory was that Kim's fans, understandably envious of her serenely untroubled life, began to emulate her *attitude*, so that they might at least *sound* equally invulnerable. The fry was thus an adaptive vocal survival mechanism like any other.

But I have come to believe that the fry's use in women has, since the presidential election of 2016 and the subsequent rise of the #MeToo movement, morphed into something quite different. The fry (which is the lowest pitch that a voice can attain) has become a way for women to level the vocal playing field with men, who (as we have seen) use their more baritone voices to dominate in conversations. The fry is only partially successful in this because it also reduces volume, which hardly helps when trying to be heard over a mansplainer—but this is perhaps mitigated by another message that is woven into our DNA. Vocal fry is produced by the same set of laryngeal muscles that nonhuman animals use when facing off against an aggressor: muscles that stiffen the vocal cords so that air from the lungs passes through them in discrete pops. Vocal fry is literally a *growl.* Not, "I am woman, hear me roar." Vocal fry is a more potent signal. Roars have a theatrical, bluffing quality. A growl says: *Watch it: I mean business.*[44]

For all the heat that young women have taken for vocal fry, the speech habit is not limited to females. One study, in the *Journal of the Acoustical Society of America* in 2016, found that it was in fact *more prevalent* in men—although not as instantly identifiable since a low, deep, nonprosodic tone seems less anomalous coming from male lips. Indeed, we expect it. *Language Log*'s Mark Liberman pointed out that even as NPR's Ira Glass reported on the storm of angry emails about his female cohosts' strong vocal fry, Glass was speaking *in a strong vocal fry*.[45] And I can think of at least one famous and successful male movie actor whose pronounced fry is critical to lending his voice the murmuring, low-pitched monotone that no critic has ever described as irritating, affected, "hesitant," or likely to "undermine . . . success." Indeed, it only bolsters the impression of unruffled cool for which he is famous. George Clooney. Since noticing this, I have started to hear the prominent fry in the voices of Matt Damon, Brad Pitt, Leonardo DiCaprio—indeed, every single popular male actor you can name.

The sexual dimorphism of the human voice clearly has a cultural component, but one that is merely layered atop a strong biological one. The producers of *Seinfeld*, to create a voice in Dan indistinguishable from that of a woman, had to post-dub a female actress's voice.[46] And people who transition to another gender often find their voices to be stubbornly stuck in their gender of origin. For transwomen (who have transitioned from male to female), estrogen treatment will effectively feminize the physical appearance (smoothing the skin, plumping the breasts) but it cannot reverse the effects of puberty on the larynx and vocal cords. Someone who transitions after puberty will in all likelihood be saddled with a voice that signals an XY, male, chromosomal makeup.

This is not a trivial consideration for transwomen who wish to successfully "pass" as biological females, nor is it trivial even for those transwomen content to project a more ambiguous, or androgynous, identity. Trans people of every shade are a population especially vulnerable to job discrimination, family rejection, and transphobic violence; and for transwomen with a sexual orientation toward men, a stubbornly male-sounding voice can rob the speaker of a major signal by which humans alert potential partners of their romantic interest, leaving some feeling dangerously isolated, or condemned to a life alone. (Not for nothing do transgender people have a suicide rate nine times that of the rest of the population.[47]) For female-to-male transmen, the situation is different, at least regarding the voice. Testosterone replacement therapy will not only masculinize a biological female's appearance (increasing muscle mass, spurring beard growth, even initiating male pattern baldness), it masculinizes the voice by binding to androgen receptors on the vocal cords and larynx, making them expand rapidly in size. The voice naturally deepens.

Surgeries for feminizing the voice in transwomen exist: one procedure permanently pulls apart the cartilages to which the opposite ends of the vocal cords attach: the increased tension on the membranes raises pitch. But the procedure often limits the cartilage's mobility, diminishing variation in prosody. People who have undergone this surgery often speak in a one-note falsetto. A more effective procedure, called endoscopic glottoplasty, reduces the amount of vibratory tissue by shortening the vocal cords. This raises pitch and retains prosody. But because operating on the delicate vocal cords carries significant risk and can even make a voice sound raspy and ragged (as Julie Andrews learned), doctors recommend such surgery only as a last resort.[48]

Most transwomen are first urged to try conventional speech therapy. This can be remarkably effective in feminizing the voice, but the techniques (which involve an ingenious manipulation of vocal auditory physics) can be difficult to master. The speaker lifts the larynx in the neck and holds it in a permanent, elevated "swallowing" position, while also subtly raising the tongue toward the palate and pushing it forward while speaking and, at the same time, pulling the lips in slightly against the teeth. By thus shrinking the resonance chambers of throat and mouth, the speaker boosts the higher frequencies in the voice spectrum and mutes the lower ones. The voice not only sounds higher, but the timbre has the lighter, brighter quality suggestive of a smaller female body. But the muscular strength and coordination necessary for permanently elevating the larynx and reshaping the oral cavity while speaking requires extensive practice. Some never get the hang of it. Meanwhile, yawns, coughs, and sudden bursts of laughter—all "fixed action patterns" that emerge from the brainstem and thus circumvent conscious control—can be treacherous: a sudden masculine sound can interrupt the female-sounding flow of speech, with jarring results.

Many trans voice tutors also warn that sounding convincingly female is not solely a matter of pitch and timbre. Whether through innate differences, or cultural conditioning, women tend to speak with a slightly different prosody than men. Feminizing the formerly male voice requires greater "flow" across speech sounds, blending the transitions between individual phonemes, and a reduction of sudden, wide pitch contrasts, to rid it of the more angular, jagged way men usually talk.[49]

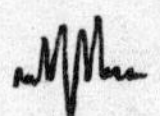

There is a third style of sexual vocal signaling in humans, distinct from the dimorphic male and female voices: this is the so-called gay voice. (If any reader somehow fails to know what I am referring to, I encourage them to listen to some of the more flamboyant stars of *Queer Eye for the Straight Guy*.) Linguists who have analyzed the features of gay voice say that it includes wider than usual swoops in pitch from high to low, exaggerated vowels (where the back of the tongue is dramatically lowered on the *ah* and *uh* sounds, often stereotyped in the utterance *faaaahb-u-luuhs*)[50] and by longer, higher-pitched hissing on *s* and *z*.[51] Not all gay men speak this way, of course, and not all men who do are gay, but the statistical correlation between these "gay voice" features and same-sex orientation is sufficiently high that, in controlled tests, listener-judges can predict with 80 percent accuracy which speakers are homosexual through voice alone.[52]

A corresponding "lesbian voice" is less well defined, acoustically. Some listeners associate a low-pitched, raspy, or assertive voice with a lesbian orientation (presumably because of the stereotypically "masculine" acoustic cues), but there's no evidence that these voice qualities are statistically higher in lesbian than in straight women. Which is why I say that there are three distinct sexual voices in humans, rather than four—although some researchers have detected a very subtle feature typical of lesbian voices, a tendency to pronounce the vowels with a slightly lowered back of the tongue, which boosts the lower frequencies in the voice spectrum.[53]

Early studies on gay and lesbian voices in the 1980s included speculation that the speech patterns and timbre differences might originate in anatomical changes in the vocal tract, or in the neurological wiring that controls articulation, brought about by the biological predispositions that produce same-sex attraction. But no anatomical differences exist in gay men (or lesbians), and no differences in their brain wiring have been found. Today, most experts agree that the voice we label as gay (or lesbian) is a learned behavior that might begin in infancy, when language is being acquired. The theory is that toddlers who are biologically predisposed to a same-sex orientation identify more closely with opposite-sex parents and caregivers, and wire in the circuits for speech patterns that conform to those speakers. Thus, a gay three-year-old boy will attend to, associate with, and eventually mimic the greater articulatory clarity and elastic prosody in his mother's voice than the more monotone, less crisply articulated, speech of his father; while a lesbian toddler associates with the father's deeper voice and thus exaggerates the lower back vowels to deepen overall voice pitch.[54] People adopt these speaking patterns unconsciously and may manifest them very early in life (some report realizing that they spoke differently than their peers in elementary school), others not until adolescence.[55]

It's easy to see why such voices evolved in our species: because same-sex attraction is not manifest in physical terms (you can't tell if someone is gay by body type, hair or skin color, facial features, and so on), the voice became a highly useful means for alerting others who share your orientation. But a voice that strongly announces same-sex attraction can also be a major trigger of discrimination. Many gay men report having been teased as small children for their "sissy" way of talking, and threatened or attacked for their voices as adults. Some internalize this homophobia and come to hate the sound of the gay voice—their own and others'. In the documentary *Do I Sound Gay?*, the filmmaker David Thorpe describes his gay friends as sounding like "a bunch of braying ninnies," and he feels his own voice stigmatizes him as a particular kind of person: a frivolous party animal, not the message he wishes to send as a thirtysomething, newly single man in search of a lasting relationship.[56] Thorpe visits a speech pathologist who counsels him to speak with less of a "singsong" prosody; to shape his vowels with his tongue more to the front of his mouth; to limit "nasality"; to reduce the hiss in his sibilants, and to stop mimicking the "upspeak" affectation typical of teenaged girls who end every statement on a rising pitch as if it's a question?

Thorpe tries hard: he monitors his tongue and lip moves in a mirror; reads aloud tongue-twisters to change his *s*'s; consults a dialogue coach to unlearn his habit of ending sentences with upspeak. But he fails; he has clearly grooved his "gay" articulations and prosody so deeply into his basal ganglia that he cannot deprogram them. Mystified at his inability to change something in himself that he dislikes, and unable to pinpoint when he even began talking in this manner, he asks family members and childhood friends when *they* noticed that he was speaking "differently." They tell him that his gay voice did not emerge until college (when he came out). By the end of the documentary, Thorpe, aware that he cannot change it, says that he has come to terms with his voice, but it is clearly an uneasy peace. Stand-up comedian Guy Branum, who, as an act of personal liberation, refuses to modify his stereotypically gay speech patterns, nevertheless understands the fear and revulsion that men like Thorpe feel for what Branum calls the "bright plumage" of their gay voice. "We are prey," Branum writes in his 2018 memoir, *My Life as a Goddess*, "and bright plumage can get you killed."[57]

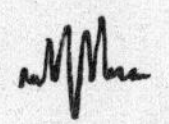

Whether you're straight, gay, lesbian, bisexual, or trans, certain acoustic changes to the voice are universal during erotic arousal. This is because the vocal tract behaves much like the sex organs—which is to say that, in the early stages of the sexual response cycle, the mucus membranes in the larynx (like those in the vagina) secrete lubricant, coating the vocal cords in sticky mucus that limits their efficiency in chopping the airstream and lends the voice a husky, breathy quality in both men and women. It also makes pitch control difficult, which is why your voice veers weirdly when you're trying to talk to someone who is making your heart race with romantic and erotic excitement. Meanwhile, swelling in the tissues of the throat and tongue (much like that of the penis and vulva) softens the walls of the vocal tract's resonance chambers, so that they absorb and muffle higher overtones, lending the voice a velvety texture. Our ears can detect, and interpret, these acoustic signals of desire—which is why, on a first date, you might be communicating more than you intend when you innocently ask for the salt.[58]

SIX

THE VOICE IN SOCIETY

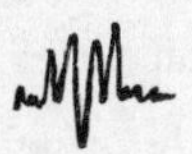

The most famous literary work about how the human voice shapes identity, and thus destiny, is George Bernard Shaw's 1912 play *Pygmalion*, later adapted into the stage musical *My Fair Lady* (starring Julie Andrews) and movie (starring Audrey Hepburn). The plot concerns a speech scientist, Henry Higgins, who transforms a penniless Cockney flower girl into "a duchess" by altering how she talks. At first, Eliza Doolittle is almost unintelligible to anyone but her fellow Cockneys. "Wal," she says in the opening scene (as rendered in Shaw's phonetic spelling), "fewd dan y'd-ooty bawmz a mather should, eed now bettern to spawl a pore gel's flahrzn than ran awy athaht pyin." (Translation: "Well, if you'd done your duty by him as a mother should, he'd know better than to spoil a poor girl's flowers then run away without paying.") After eight weeks of phonetic training (and some adjustments to her dress and grooming), Professor Higgins has made a new woman of Eliza. She speaks in an impeccable upper-class accent, attends society garden parties, is preparing to launch her own phonetics practice—and is about to marry into the aristocracy.

Shaw, an avid socialist, saw *Pygmalion* as a critique of England's rigid class system and a commentary on how accents trap people in the social stratum into which they happen to have been born. He put it bluntly in the play's preface: "it is impossible for an Englishman to open his mouth without making some other Englishman hate or despise him."[1] For this reason, Shaw formed the unusual belief that phoneticians, like Higgins, were the most important social reformers in England. They alone could rid society of the vocal differences that so stubbornly divide the classes, and that deny people like Eliza any social mobility.

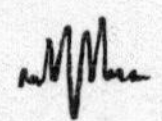

Everyone has an accent—a fact about which you can remain ignorant only if you never encounter someone who speaks your shared language differently than you do. Just as every human speaker emits vocal signals whose shape, rhythm, and tune offer strong clues about their geographic origins, socioeconomic background, and education level, all of us, as listeners, parse other people's pronunciation for such clues and draw instant inferences, often quite inaccurate, about the speaker. Accordingly, some linguists call voice and accent the last socially acceptable form of prejudice.

It is also a "prejudice" over which we, as listeners, have little or no control. All stimuli—visual, tactile, auditory—are first processed in our emotional limbic system, which parses them for threats. Only then is the signal passed up to our higher brain for more rational analysis. So even before we have considered *what* someone is saying, we experience certain instinctual, limbic reactions to *how* they're saying it, including reactions to accent. In 2015, psychologist Patricia Bestelmeyer of Bangor University used fMRI to show that activity in our limbic system, including the hot-button amygdala, determines whether a speaker is, according to his or her accent, a member of our "in-group," or an outsider, an "other."[2] Bestelmeyer theorized that such reactions date to when speech first evolved, as a way for early humans to assess whether someone belonged to the same tribe as the listener, and was friend or foe—which suggests that there was more neuroscience in Shaw's comment about accents triggering hatred than is immediately obvious. But even if violent emotional reactions to accent are, to a degree, hardwired, we are also rational, evolved creatures capable of overriding atavistic responses (otherwise, we would still be behaving like our less civilized cousins, the apes)—and Shaw rightly condemned the injustices that vocal prejudices give rise to. In *Pygmalion*, he took pains to show that Eliza, before her vocal transformation, possesses an innate decency, appetite for hard work, and determination to better herself that make her at least the moral equal of the aristocratic Londoners who turn their noses up at the sight, and perhaps especially the *sound*, of her.

Recent studies show that, more than a hundred years after Shaw wrote *Pygmalion*, little has changed: people in the United Kingdom still associate the educated upper-middle-class accent that Higgins teaches Eliza with honesty, intelligence, ambition, and even physical attractiveness,[3] while her Cockney accent is associated with lack of professionalism, low prestige, unintelligibility, and lack of success.[4] That some ways of talking speed professional advancement, while others relegate entire segments of the population to menial labor, the welfare rolls, or worse, is a remarkable aspect of our vocal signaling. The voice, in other words, drives our *social* evolution as a species, helping to structure human civilizations. English is ideal for studying the phenomenon not only because it has become the global language of business and diplomacy, but because the place where it first emerged also happens to be the place where you can still find the greatest variety of accents and the greatest class snobbery associated with them. But speakers of every language make snap judgments about others according to voice and accent, and English speakers in New Zealand, Australia, South Africa, and the United States are (as we will see) no exception.

English, as it developed in the British Isles over the last 1,500 years, was spoken by people from Wales to London, Birmingham to Dublin, Manchester to Edinburgh, from remote rural redoubts, to large urban centers, and by people of widely differing levels of education, occupation, and wealth. Accents emerge in groups of people who are isolated or "islanded"—geographically, financially, socially, professionally—so that a particular way of pronouncing speech sounds becomes, in effect, "inbred" by being passed down from caregivers to children by Motherese, over generations. Inevitably, many different ways of sculpting an *a* or an *e*, or tapping out a *t*, emerged across England.

By Shakespeare's time (in the late sixteenth and early seventeenth centuries), the accent associated with educated Londoners had already taken on a certain prestige, since the city was the center of finance, art, and politics, and the way people did things there tended to have an outsized influence on how they did them everywhere else.[5] But a remarkable number of other accents flourished across England and still do. In his book *The Mother Tongue* (1990), Bill Bryson notes that in "six counties of northern England, an area about the size of Maine, there are seventeen separate pronunciations of the word *house*."[6] All told, there are some one hundred distinct accents across Great Britain—a remarkable total for so small a geographic landmass.

Efforts to reduce English to a single "standard" pronunciation began in earnest in the mid-eighteenth century when Thomas Sheridan, an ex-actor and teacher, appointed himself *the* authority on how English should sound, and became the country's first National Elocutionist. The subtitle of his bestselling book *British Education*, published in 1756,[7] left no doubt about the connection Sheridan drew between speech patterns and moral character:

> Being an Essay towards proving that the Immorality, Ignorance, and false Taste, which so generally prevail, are the natural and necessary Consequences of the present defective System of Education. With an attempt to shew, that a revival of the Art of Speaking, and the Study of Our Own Language, might contribute, in a great measure, to the Cure of those Evils.

Sheridan aimed "to fix and preserve [English] in its state of perfection"—which is to say, how well-educated, well-heeled, and well-connected Londoners spoke it. He denounced the Cockney habits of dropping the *h* in words like "Heaven" and "happy," and pronouncing *th* as *f* or *v* (as in, "My bruvver finks 'e's in 'eaven"). He derided speakers from northern cities like Liverpool for pronouncing the *u* in "cup" with dropped tongue and rounded lips that made the word sound like "coop"; he scorned the tendency of the Irish to "mispronounce" the vowels *o* and *e* (so that "sort" became "s*a*rt" and "person" became "p*ai*rson"), and he informed Scottish-accented speakers that they were doing almost everything wrong. The book was popular among the country's growing middle class, whose insecurities about gaining membership in polite society Sheridan preyed upon mercilessly. His follow-up volume, *A Course of Lectures on Elocution*, announced that speaking like an educated Londoner (or member of the royal court) is "a sort of proof that a person has kept good company, and on that account is sought after by all, who wish to be considered as fashionable people, or members of the beau monde."[8] All other accents, he warned, "have some degree of disgrace annexed to them."[9]

As snobby as Sheridan can sound, his stated aim was the opposite of accentuating class and regional divisions. He imagined (like Shaw, a century and a half later) that by erasing differences in how English was spoken, all Britons would become mutually intelligible to one another, fostering greater communication between *all* people. A fine aim and with fine logic behind it. If our evolutionary history predisposes us to fearful reactions to accents not our own, then eliminating differences in pronunciation could only be a good thing. On a practical level, reducing accents to a single standard would abolish all kinds of miscommunication, as when the pronunciation that, say, a Cockney speaker gives to the word "pie" might seriously confuse an upper-class Englishman who hears him state a desire to eat some "poy." Differences in consonant pronunciation also create confusion: a Cockney fruit seller who says that he has "three apples" could be accused of dishonesty by the Oxford graduate who has *clearly* heard him say "*free* apples."

But despite his bestselling books and a national lecture tour that packed auditoriums, Sheridan failed to convince everyone to talk as he did. Most people in Scotland and Ireland proved perfectly happy with how they sounded—proud of it, in fact—and ignored Sheridan's rules for "fashionable" pronunciation. Others (perhaps too poor to afford Sheridan's books, or too busy to attend his lectures, or simply lacking the articulatory prowess) were stuck with their stigmatizing Cockney brogues, Yorkshire burrs, or other regional signatures. So, ironically, Sheridan's effort to unite the country only drove a further wedge between people, by drawing attention to the supposed "disgrace" attached to specific accents, and holding up one particular form of speech as the only correct one.[10]

This vocal divide took on truly toxic class dimensions in the nineteenth century when the Industrial Revolution brought people from all corners of the country pouring into London and other urban centers to work in the factories that were creating the greatest explosion of wealth and prosperity the world had ever seen—at least, for the factory *owners*. The uneducated, underpaid factory *workers* (many of them children under the age of eight), were consigned to dire Dickensian slums, where the dominant accent (Cockney) became associated with all the ills of poverty: prostitution, alcoholism, robbery, murder. Meanwhile, the upper middle class—some of them former "rustics," provincials, and even Cockneys who defied the odds and clawed their way up the income ladder—scrambled to enroll their children in a new type of school: high-tuition preparatory and boarding schools, where one of the chief attainments was an accent that bore precisely no embarrassing giveaway traces of their parent's "disgraceful" origins.[11]

The accent was a version of the one promoted by Sheridan and spoken by children at the country's top prep schools (Eton and Harrow) and universities (Oxford and Cambridge). Known variously as "correct English," "good English," "pure English," "standard English,"[12] it eventually acquired the name "Received Pronunciation," or RP, and its chief distinction was that it had no distinction at all—it deliberately erased all signs of regional accent, and thus all indications of what slum or hamlet you came from. It remains, to this day, the prestige accent of Britain's educated upper middle class, and the target accent taught to foreign students of English abroad. Although spoken by only about 3 percent of the British population, it's a prominent and powerful 3 percent. Anyone who has ever heard former prime minister David Cameron speak, or watched *Downton Abbey*, or seen a Hugh Grant movie, knows RP: its features include the elongated *a* that makes the word "bath" take on the posh-sounding pronunciation "baaawwth"; it removes the *r* sound from the ends of words and syllables so that "purple" and "learn" and "more" become "puh-ple,""luuhn," and "mo-ah"; it mandates that the lips be held in against the teeth when saying things like "cup" so that the word doesn't shade toward "coop," and it calls for the studious aspiration of *h* at the beginning of words. (A still more tony version of RP, called Refined RP, is reserved almost exclusively for the royal family and their "noble" retinue, and is responsible for pronunciations like "bleck hit" for "black hat" and "Ehh lewktool ayvah th' hice," for "I looked all over the house.")

Thanks to the plasticity of the preadolescent linguistic brain, a year's exposure to this approved, standardized accent was usually sufficient for Victorian public (that is, private) school children to rinse the shaming strains of Yorkshire, the Midlands, or, God forbid, Cockney, from their voices. This accent leveling was imposed by schoolmasters (who, in rebuking students for saying "loike" for "like," exhorted them to "Keep your *i*'s pure!"[13]). But the accent was also ruthlessly enforced by peers. In the 1986 book *The Story of English,* Robert MacNeil and his coauthors quote from a history of Victorian private schools by John Honey, who documented how, at one such school, in the Midlands of the 1800s, "local boys with a North Bedfordshire accent were so mercilessly imitated and laughed at that, if they had any intelligence, they were soon able to speak standard English."[14] Those who couldn't were stigmatized as undereducated, ignorant, provincial, poor—the out-group.[15]

Thus did England, by the dawn of the twentieth century, make voice one of the primary means by which people categorized one another.

RP, Received Pronunciation, was the accent Henry Higgins taught to Eliza Doolittle and he did so with a then new science of pronunciation, *phonetics*, that studies the exact movements of the larynx, tongue, velum, and lips in speech. Charles Darwin's grandfather Erasmus (1731–1802) made the first phonetic study when he catalogued the precise tongue positions for vowels by placing cylinders of rolled tinfoil in his mouth, noting where his tongue left indentations. But the father of modern phonetics was Alexander Melville Bell, whose son, Alexander Graham Bell, would go on to create the most revolutionary invention in human voice communication, the telephone. Bell *père*'s work was, in its way, as lasting. In the mid-1800s, he created a unique set of hieroglyph-like symbols that represented every possible position of the articulators and thus every sound that could be generated by the human vocal tract, in any language, including the glottal clicks and tongue pops of African tongues like Xhosa. Bell tried to market his findings in a privately printed book, *Visible Speech* (1867), which he touted as an aid to foreign language learning. It failed to sell. But as the first accurate and exhaustive speech-transcription system, Bell's *Visible Speech* would have serious legs. A streamlined version was produced by one of Bell's students, the phonetician Henry Sweet (upon whom Shaw based the fictional Henry Higgins). Sweet replaced Bell's weird hieroglyphs with familiar Roman letters (capitalized, italicized, turned, or raised), and with a team of other phoneticians from around the world created the International Phonetic Alphabet, a system still in wide use today and which serves as the pronunciation guide in every dictionary (or dik-shə-ner-ē) and is an indispensable tool for lexicographers, foreign language students, speech pathologists, translators, even singers and actors—and which George Bernard Shaw believed, when he was writing *Pygmalion*, would change the world as a tool by which class differences in human speech could be erased. That dream became moot when a (seemingly) far more effective means for influencing English pronunciation arrived on the scene.

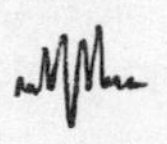

The BBC made its debut radio broadcast on November 14, 1922, a decade after *Pygmalion*'s premiere. That broadcast marked the first time that a single voice addressed an entire nation. It did so in flawless RP. This was no coincidence. The network's administrators and announcers (all drawn from Britain's ruling elite) had mandated that RP be the BBC's exclusive accent, arguing that it was the clearest, most intelligible style of speech (which is, of course, what speakers of *any* accent say about their own way of talking), but also the most inherently "beautiful."[16] By broadcasting exclusively RP voices, the BBC also hoped that it could succeed where Sheridan and Shaw had failed: transforming the speech habits of the nation by leveling the accent to a single standard—a hope openly stated by the corporation's first director-general, Lord Reith. "One hears the most appalling travesties of vowel production," he wrote soon after the BBC's maiden broadcast. "This is a matter in which broadcasting may be of immense assistance."[17]

To that end, Reith, in 1926, created the BBC's Advisory Committee on Spoken English, whose mission was to police on-air accents and to settle disputes about tricky pronunciations (should the Latinate word "canine" be *kah*-nine, or *kay*-nine?). George Bernard Shaw became a committee member and served for ten years, eventually rising to chairman—despite a distinct Irish accent acquired during his childhood and youth in Dublin. Shaw, however, was well versed in RP—and may, indeed, have submitted to an Eliza Doolittle–like course of phonetic instruction in his early twenties[18] upon arriving in London as a penniless, tongue-tied, high school dropout; recordings made of him in the 1920s, when Shaw was in his seventies, reveal a hybrid accent that mixes his Irish lilt with some curiously RP-sounding vowels. Whatever the case, the majority of Shaw's fellow committee members all spoke in the toniest, Oxford- and Cambridge-honed RP, including art historian Kenneth Clarke (who later hosted PBS's *Civilization*) and journalist Alistair Cooke (later host of PBS's *Masterpiece Theatre*).[19] Evidence suggests that *their* pronunciations tended to prevail in internal debates of the Advisory Committee. (For instance, Shaw's argument for "*kay*-nine" was rejected as an uncouth "Americanism.")[20]

The BBC, of course, failed utterly to "level" accents across Great Britain. And with good reason. As noted earlier, there is a critical childhood period for learning accents, so for all adult listeners to the BBC, the announcers' RP accents were powerless to change how they spoke. Like the light bulb in the joke about psychiatrists,* people who change their accent after puberty *really have to want to change*—by submitting to the kind of intensive training and practice that Eliza (and possibly Shaw) underwent. But babies and small children also proved entirely immune to the accent-leveling influence of the BBC. Disembodied electronic voices are powerless to inculcate babies with language or accents: speech can only be acquired as part of the back-and-forth feedback loop of a caregiver speaking Motherese and a baby babbling in reply.

The BBC's Advisory Committee was shut down in 1939, as an unnecessary expense in wartime. It was not resumed after the war—but not for budgetary reasons. In 1945, the country took a hard-left turn, politically, electing its first majority Labour government. With a newly empowered working class in charge, attitudes toward RP, and the 3 percent of people who spoke it, shifted dramatically. It was suddenly no longer so fashionable or desirable to sound like a member of Britain's posh ruling class, and the dream (shared by Sheridan, Shaw, and the BBC) of leveling all

*__Q: How many psychiatrists does it take to change a light bulb?__
A: Just one—but the bulb really has to want to change.

speech to a single (upper-class) accent was perceived, not as a benevolent way to erase class differences, but as a patronizing imposition, by a privileged minority, on the culture and identity of the majority working class. Daniel Jones, the phonetician who had coined the term Received Pronunciation, hastened to distance himself from the whole concept. In the 1944 edition of his *English Pronouncing Dictionary*, he insisted that he did "not regard RP as intrinsically 'better' or more 'beautiful' than any other form of pronunciation. . . . I take the view that people should be allowed to speak as they like."[21]

The BBC apparently now agreed because it soon hired some on-air talent with regionally tinged accents. By 1962, the widespread acceptance of non-RP speech over the British airwaves was such that, when the Beatles arrived in London on the first wave of Beatlemania, they were able to ignore the practice of earlier Liverpudlian entertainers who, upon making it in the capital, were required to jettison their Liverpool "Scouse" for a plummy RP. Indeed, daring to speak in their natural voices was an act as revolutionary as the Beatles' eyebrow-brushing bangs, narrow suits, melodic innovations, and Cuban-heeled boots—and it helped (at least for a while) to make the Liverpool accent, and indeed all working-class voices, downright fashionable in 1960s Britain. RP speakers of long standing quickly learned to "code-switch" (that is, to alter their accent for purposes of social advancement—to join the in-group), as when the Rolling Stones' front man, Mick Jagger, a product of solidly middle-class Dartford, in Kent, and a former student at the London School of Economics, began speaking in a caricatured Cockney. Except when he happened to find himself, in May 1966, on a talk show defending himself against criminal charges of marijuana possession after a drug bust. Jagger reverted to his best boarding-school RP. The members of Monty Python's Flying Circus deployed *their* upper-crust Cambridge accents to satiric effect through preposterous exaggeration of its vowels and consonants (to say nothing of their absurdist utterances), and the comic bestselling books *Fraffly Well Spoken* (1968) and *Fraffly Suite* (1969) parodied, through clever phonetic spelling, the ridiculous, lockjawed sound of Refined RP. "Bar chorleh a smol gront from the yacht skonsul snommotch twosk for" (Translation: "But surely a small grant from the arts council is not much to ask for").[22] The *Fraffly* (or "frightfully") *Well Spoken* series was by Australian author Alastair Ardoch Morrison writing under the pseudonym Afferbeck Lauder (Refined RP for "Alphabetical Order"). No professional linguist has ever demonstrated so potently how even the most supposedly "refined" accent is intelligible only to those who happen to share it.

Accents are, however, like hemlines: fashionable pronunciations rise or fall in popularity and prestige according to the whims of social and political change. With the election of Margaret Thatcher's Tory government in 1979, RP was back in (soon giving rise to a mania, in the early to mid-1980s, for the Sloane Ranger accent popularized by Lady Diana Spencer, then-fiancée to Britain's *Fraffly*-speaking Prince Chulz). Today, in studies designed to rank the most desirable British accents, RP is once again the sought-after voice, while the Liverpool accent, after its brief, Beatles-inspired moment of desirability, has been thrust back to the bottom of the social ladder. Along with Cockney (polls show), Scouse is the accent most likely to ruin your chances for landing the big job, to be accepted in your proposal of marriage, or to win an election.

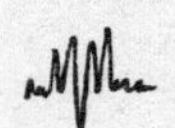

George Bernard Shaw (or maybe Oscar Wilde—no one seems to know for sure) once quipped that America and Great Britain are "two countries separated by a common language." There are indeed enough vocabulary and pronunciation differences to make the two tongues seem, at times, distinct. But even so, they operate identically on the level of voice—which is to say that Americans are as quick as the most accent-conscious Briton to form snap judgments about their fellow countrymen and -women according to how they move their lips and tongue when speaking English. There are fewer accents across the United States than across the United Kingdom (about one-fifth, in fact), but that's only because England had a thousand-year head start in shaping its various dialects. There are, in any case, more than enough different accents across the United States over which to discriminate—such that you might almost say that an American can't open his mouth without making some other American hate or despise him.

Thus, to many Americans dwelling above the Mason-Dixon line today, something as arbitrary as a Southern speaker's pronouncing the *i* vowel as *ah* ("Ah lahk pah" for "I like pie") can instantly stigmatize the speaker as backward, undereducated, slow-witted, prejudiced, or intolerant; while to Southerners, a Northerner's habit of pronouncing *i* as two vowels (the dipthong *uhh-ee*) establishes the speaker as an elitist, PC-liberal snob. Midwesterners, with their hard *r*'s, nasality, and singsong prosody, are, to East Coasters and Southerners alike, bumpkins (like Marge Gunderson in the Coen brothers' *Fargo*), while Californians, with their surfer-dude drawls and Valley Girl upspeak, are, to the rest of the country, hopeless flakes.

Americans also make vocal judgments about a speaker's economic status as readily and automatically as do the British. No American who reads *The Great Gatsby* wonders what the title character could possibly be referring to when he says of Daisy Buchanan: "Her voice is full of money."[23] The "money" Gatsby hears is partly a feature of what she says (a blend of weightless gossip and empty flirtation possible only in someone with plenty of leisure time), but mostly *how* she says it—not only her accent, but her prosody, pace, and timbre. "She began to ask me questions in her low, thrilling voice," the novel's narrator, Nick Carraway, rhapsodizes: "It was the kind of voice that the ear follows up and down as if each speech is an arrangement of notes that will never be played again . . . there was an excitement in her voice . . . a singing compulsion, a whispered 'Listen,' a promise that she had done gay, exciting things just a while since . . ."[24]

Fitzgerald filled *The Great Gatsby* with his abiding obsessions over money and class, and their manifestations in the voice, in part because he wrote the book in the early 1920s, alongside the birth of radio. Alive to every thrilling new innovation of the Jazz Age, he seemed determined to catch the vividness of his characters' voices with all the immediacy of a radio play, and the novel is (to my knowledge) unique in American literature for its focus on how people *sound.* It is virtually a book about voices. Daisy's, for instance, proves to be a "dishonest signal." For all the warmth and intimacy her voice projects, *the fun*, she turns out to be as cold and opportunistically self-dealing as her odious husband, Tom, upon whose voice Fitzgerald lavishes remarkable attention, capturing not only the dominance and imperiousness typical of his class, but its forced masculinity:

> His speaking voice, a gruff husky tenor, added to the impression of fractiousness he conveyed. There was a touch of paternal contempt in it . . . "Now, don't think my opinion on these matters is final," he seemed to say, "just because I'm more of a man than you are."[25]

Meanwhile, poor Gatsby, the social-climbing arriviste, has a voice whose pretentions to high-born status are captured in his clumsy, would-be "British" expression: "Old sport." Only at the end of the book, when Gatsby's father (a farmer from rural Minnesota) arrives in Long Island for his son's funeral do we learn how the young "James Gatz" would have sounded before the self-improvement program he subjected himself to as a boy, an Eliza Doolittle–like regimen that included an hour a day (5:00 to 6:00 p.m., his childhood planner reveals) to "Practice elocution, poise and how to attain it."

The linguist who has provided the greatest understanding of how American voices stratify the country along socioeconomic lines is William Labov, an emeritus professor at the University of Pennsylvania. At age ninety-two, Labov enjoys a reputation among linguists equal to that of his contemporary Noam Chomsky. But whereas Chomsky focuses on everything in language that doesn't change (the supposed "universals"), Labov's obsession, for the last six decades, has been everything that *does*,[26] with a special emphasis on accent—an approach that has given rise to an entire subspecialty of voice science called sociolinguistics.

Labov's first foray into this undiscovered realm came with his 1960 Columbia University master's thesis on a subtle sound change in the voices of the year-round residents of Martha's Vineyard, the tiny island of six thousand inhabitants off the coast of Massachusetts.[27] Through skillful interviews, Labov learned how the local fishing families' pronunciation of certain vowels (which went back to when their ancestors settled the island in 1642) was an unconscious expression of the loathing they felt toward the summer people—the wealthy professionals from mainland cities like Boston and New York who descended on the island every July and August. The culture clash between natives and summer people was exacerbated by the recent collapse of the local fishing industry, which had made the fishing families of the north shore the poorest people in the state, with the highest unemployment. This, in turn, had allowed the summer people to buy up, at fire sale prices, all the large oceanfront houses built by the original settlers, driving their descendants from their ancestral homes into the hills and hollows of the interior. The fishermen's reverting to an old accent (a study from thirty years earlier showed that the families used to pronounce the vowels very much like people on the mainland) was, Labov said, an unconscious attempt to keep alive the vanished glory of the Vineyard's storied fishing and whaling past, to express their sense of themselves as the true "owners" of the Vineyard (accent as territorial marking), and to distinguish themselves from the hated mainland interlopers and their out-group big-city vowels.

Linguists had long known that language and accents are in constant flux, but Labov's Vineyard study was the first to document a sound change *in action*—and to describe the social pressures driving it. The study provided a template by which Labov, over the next five decades, laid bare the socioeconomic determinants of various subtle (and not so subtle) accent changes across the country, including the sudden pronunciation, by upper-middle-class Manhattanites, in the mid-twentieth century, of the hard *r* sound after vowels (in words like "ca*R*," and New Yo*R*k"), to distinguish themselves from city dwellers of working-class origins, who said "cah" and "New Yawk."[28] Today, Woody Allen and Bernie Sanders are the most famous bearers of this working-class New York accent feature. But seventy years ago, even rich, "aristocratic" New Yorkers, like President Franklin Delano Roosevelt, left off the *r* after vowels. (In his inaugural address of 1933, he can be heard declaiming: "We have nothing to *feah* but *feah* itself.")

This missing *r* (also heard in Massachusetts, where nonnative actors trying to portray Bostonians practice it with the elocution exercise "Pak yah cah in Hahvahd Yahd," and in the South where Virginians and Carolinians eat "po' boys" and vote for "senatahs") is a linguistic legacy of the original British settlers of the seventeenth century (and is, of course, a major feature of the RP accent of Hugh Grant and the *Downton Abbey* cast today). But if that *r*-lessness was part of the linguistic umbilical cord that still connected cities of America's eastern seaboard with the motherland, that cord was severed in some cities after the Second World War, when the United States came of age as the world's superpower. Wealthy New Yorkers, in particular, unconsciously began to distinguish themselves from both a declining England, and the un-moneyed Manhattan masses, by wrapping their well-bred lips and tongues around those terminal *r*'s. Labov documented the change, as it was occurring, in an ingenious experiment for his 1962 PhD dissertation, by plotting the statistical prevalence of *r* in the voices of clerks in three Manhattan department stores: high-end Saks, midrange Macy's, and super-discount S. Klein. By requesting an item that he had previously established was on the "fou*r*th floo*r*," Labov was able to show that the statistical distribution of the *r* sound in New York City was directly proportional to the social class to which the speakers belonged (or wanted to be *perceived* to belong).

Labov expanded his dissertation into a five-hundred-page book, *The Social Stratification of English in New York City* (1966), which examined several other class-based city accent features (including the tendency, later noted by *Saturday Night Live*'s Mike Myers, for outer borough speakers to pronounce "coffee talk" as "coo-awfee too-awk," and to drop the *g* off the suffix *–ing*, as in Ratso Rizzo's famous "I'm woo-awkin' heah!" from *Midnight Cowboy*). The book instantly made Labov famous in linguistic circles. But his most startling revelations about American speech came a few years later when he began publishing papers on a previously undocumented peculiarity in pronunciation in a vast, but puzzlingly self-contained, swath of the country, extending from upper New York state, across the Great Lakes region, to the western edge of Illinois.

Until Labov's revelations, this large expanse of the Midwest had always been understood to typify a style of speech that linguists call General American, or GA, an accent devoid of any overt regional features (like Ratso's New Yawk patois or the *r*-less sound of Mark Wahlberg's tough-guy South Boston voice, or the twang of a Texan's vowels). GA's neutrality goes back to the settlers who originally pushed west into the American interior. They were mostly from Scotland and Ireland, both of which chose not to adopt Thomas Sheridan's "fashionable" London-based speech and kept the *r* after vowels and eschewed the lip pout that turns words like "dance" into "dawnce." Which is why, when a Midwesterner today says "car," she shows no inclination to extend the vowel into a louche upper-class drawl and she wraps her tongue firmly around the terminal *r*.[29]

GA is (phoneticians are fond of pointing out) the closest America has to a "standard" accent like Britain's RP: that is, one that doesn't divulge where you came from (beyond the vast expanse of what the coastal elites call "fly-over country"). But unlike RP (a "prestige" voice of Britain's educated upper middle class), GA is sonic democracy: the ideal of "class-free" America made audible; the speech style of the country's Great Middle, both geographically and economically. Little surprise, then, that when radio and television came along, GA was made the default accent (or nonaccent) of national broadcasting—and remains so for all American-born on-air network television personalities. Don Lemon on CNN, Chris Hayes on MSNBC, Stephen Colbert on CBS—all were born or raised where strong regional accents resounded (Baton Rouge, the Bronx, and South Carolina, respectively) but all today are careful to speak in generic GA (both Lemon and Colbert have talked about the training they underwent to lose their regional accents).

It's maybe worth remarking, though, that the tacit imposition of the GA accent by the American radio and TV networks stems from a very different impulse than the one that compelled Lord Reith to mandate the RP accent at the BBC. It's impossible to imagine any American TV executive concerned about "leveling" accents across the United States. Instead, the motivation for universal GA over the airwaves is purely capitalistic. By broadcasting voices with the fewest "sectional peculiarities" (as one American elocution manual from the 1920s described regional accent markers),[30] the network bosses are simply trying to avoid the instinctive fear and disgust response to out-group vocal sounds that might compel a viewer to change the channel.

That people in the huge expanse of the Great Lakes region *do not* speak in GA (as previously believed) was a fact upon which Labov stumbled in the late 1960s while listening to a tape of a Chicago teenager named Tony.[31] Puzzled at Tony's nonsensical claim that his friend Marty nearly died "in the lax," Labov eventually realized that Tony was saying his friend almost drowned while swimming "in the *locks*"—the narrow canals leading into Lake Michigan. Further investigation revealed that Tony said all his short *o*'s as if they were short *a*'s and that, amazingly, so did people in all the major cities of the Great Lakes region—including Syracuse, Rochester, Buffalo, Detroit, Cleveland, Chicago, and Milwaukee. In all these cities, people said "I *gat* the *jab*," instead of "I got the job," and "a pair of *sax*" instead of "a pair of socks."

Labov discovered that this displacement of the low-back short *o* vowel into the high-front short *a* space in the mouth had initiated a domino-like effect through the oral cavity, pushing each vowel sound one unit forward, such that city-dwellers across the entire 350-mile-long section of the country (known as the Inland North) had begun saying words like "fat" and "Anne" like "*fiat*" and "*Ian*," and words like "boss" and "caught" like "*bus*" and "*cut*." *Saturday Night Live* writer Robert Smigel (a native New Yorker) was struck by the unusual accent when he moved to Chicago in the early 1980s, and later satirized it in a series of sketches from the early 1990s about Chicago Bears football fans. Seated around a table heaped with meats, they talk about "pork *chaps*," proclaim themselves Bears "*fee-yans*," and eagerly await the big "kick *aff*."

Labov labeled the phenomenon the Northern Cities Shift—and it's a bigger deal than it might seem. As Edward McClelland puts it in his book *How to Speak Midwestern* (2016), "Vowels whose pronunciations had been stable for a thousand years, since the days of feudal England, began taking on new inflections in the mouths of Upper Midwesterners."[32] When Labov brought in a linguistic team from outside the vowel shift region to investigate the phenomenon, the transplanted linguists found that they often could not understand what the locals were saying. One heard a radio announcer in Chicago warn that an expressway was "jammed salad"; another that a local plant could not maintain "abberations" (that is, *operations*). A hotel concierge in Detroit announced that coffee would be served each morning next to the "padded plant." Still more remarkable, however, was Labov's discovery that the vowel shift was rendering the speakers incomprehensible to *themselves.* When he played a tape of the isolated word "block," as spoken by a Chicagoan, to other Chicagoans, they identified it as the word "black"—until they heard the full sentence ("Senior citizens living on one block"), and realized their error. "Most spectacular," Labov wrote, was the locals' mishearing of the word "buses." All of them heard the word "bosses."

Given the extremity of the pronunciation changes occurring in the Great Lakes region, Labov was further startled to learn that it was extremely new. Consulting earlier phonetic descriptions of how people in the region spoke (conducted in the 1930s), Labov realized that the Northern Cities Shift had come to full fruition in *only the last fifty years*, around the mid-1960s. Why it happened tells a remarkable story of how even the most fine-grained adjustments to the movements of the tongue and lips in speech are driven by political, ideological, and cultural forces.

In research done in the early 2000s, Labov traced the origins of the Northern Cities Shift to 1817, with the construction of the Erie Canal, which opened a waterway from New York into the American interior.[33] By far the biggest migration west along the canal's route were those

RP-eschewing Scottish and Irish settlers, many of whom had been living for several generations in upper New York state and had forged an identity that today we call Puritan Yankee, a devoutly religious population strongly opposed to the death penalty and slavery. Which meant that when they arrived in huge numbers in the Great Lakes region they experienced a considerable culture clash with another group of settlers who had started moving to the area: transplanted Appalachians from the proslavery, pro-capital-punishment states of Virginia, Alabama, Mississippi, Georgia, and Kentucky. Contemporary historians noted the violence of the cultural and ideological collision: the Puritan Yankees "thought of the Southerner as a lean, lank, lazy creature, burrowing in a hut, and rioting in whiskey, dirt and ignorance."[34] The Southerner thought of the Yankee (who agitated not only for the end of slavery and capital punishment, but also espoused health foods, prison reform, women's rights, new standards of dress, the closing of businesses on Sundays, and the shuttering of saloons) as a kind of proto-parody of today's self-righteously "woke," progressive, PC-policing millennial.

Like the fishing families on Martha's Vineyard, whose loathing of the summer interlopers prompted an unconscious change in vowels to mark an in-group/out-group divide, the Northerners began to use their way of shaping the short *a* to broadcast the greatest possible ideological and cultural distance from the upland Southerners, who showed zero inclination to adopt the accent change (and, indeed, began to unconsciously push the *a* vowel farther back in the mouth, stretching it out into a nice molasses drawl). As the ideological schism deepened over the next one hundred years, both sides exaggerated the speech features that marked them as distinct from the hated other. The two accents, as Labov put it, did not "drift" apart. They were actively "pushed apart," the Northerners rotating the vowels forward through the mouth as the Southerners dropped them back. This dynamic continued with the violent upheaval of the Civil War, and was accelerated through the twentieth century, with the passage of the Civil Rights Act of 1964—to which Republican politicians, including Barry Goldwater and Richard Nixon, cynically responded with the "Southern strategy," a shameless appeal to racism against African Americans that ultimately won the Republicans the political allegiance of Southern states that, for over one hundred years, had been a Democratic stronghold. (Meanwhile, the Inland North states, formerly rock-ribbed Republican, flipped Democratic.)

Various forms of race-based fear-mongering, and culture-war wedge issues, have kept the Southern states Republican since then and reinforced the unusually sharp boundary line between the two accents. Remarkably, people in cities of the Midland states just to the south of the Great Lakes region—including Columbus, Ohio; Indianapolis, Indiana; Kansas City, Missouri; Omaha, Nebraska—show no sign of the vowel shift, and for this reason speak in an accent that can safely be described as classic General American. These GA speakers also mark a political boundary line that (at least up until the presidential election of 2016) clearly demarcates the Northern (Democratic-voting), and the Southern (Republican-voting) states as clearly as the pronunciation of "socks" and "sacks."

Not everyone in the Inland North adopted the vowel shift, however. Conspicuously missing were those people over whose fate the Civil War was fought: African Americans. To this day, despite having lived among the vowel-shifting whites of the Inland North for generations, Black residents of Buffalo, Rochester, Syracuse, Detroit, Chicago, Cleveland, and Milwaukee show no tendency toward saying "fat" like "*fiat*" or "chopsticks" like "*chap*sticks." Instead, they sound a lot like the Black populations in the rural South from which they, or their ancestors, migrated.

They also sound surprisingly like the Black populations of places many miles away: in New York, Boston, San Francisco, or Los Angeles.[35]

The accent they share includes the Southern pronunciation of long *i* as *ah* (which white people in the South do), but also the uniquely African American tendency to give the *l* sound a vowel-like quality at the end of words ("coo'" instead of "cool"), and dropping the *r* after vowels not only at the ends of words (like white, working-class New Yorkers and all Bostonians and Carolinians) but in the middle, as in "Flo'ida" for "Florida."[36] Along with these, and other, accent details come an array of grammatical features that are distinct to inner-city African American speech, including deletion of the copula verb in certain constructions (as in "We happy"—but not in the first person singular "I *am* happy"), and a use of the verb "to be" in statements such as "She be working," which indicates an ongoing state of employment stretching over years (as opposed to the observation that she happens to be at work *right now*: "She working").

Not all African Americans speak this dialect, of course, but every African American who grew up in a predominantly Black neighborhood has myelinated the necessary speech-motor circuitry for slipping into it should they choose—code-switching like Mick Jagger when shuttling between his prep-school RP and his rock 'n' roll Cockney. Linguistics professor John Baugh grew up in inner-city Los Angeles and for the most part speaks in the Standard English he heard at home (his mom was a college-educated elementary school teacher). But when greeting fellow African Americans, Baugh will often briefly drop this "formal" speech and execute a kind of verbal fist-bump, a quick smattering of the African American dialect that he grew up hearing on the playground and in the street. "So if several Black professors get together," Baugh told me recently, "it's very common to say, 'Hey, bro, what's *hap*penin'?' 'Ain't *noth*in' goin' on'—we style shift for a second. It provides a lot of valuable ethnic, linguistic, and racial solidarity when people engage in that."[37]

In the 1990s, Baugh used his talent for code-switching to conduct an eye-opening, and now classic, study of voice profiling by landlords around Stanford University. By phoning in responses to advertised apartments in expensive neighborhoods of Palo Alto, and variously using his impeccable African American "inner-city" voice, his flawless Chicano voice, and his perfectly modulated Standard English "professor" voice, Baugh revealed that landlords were claiming *no vacancy* to everyone except those who sounded white.[38]

Today, the African American dialect that Baugh code-switches into when greeting fellow Black professors, and that he used in his study of linguistic profiling, is known, variously, as "Black English," "African American Vernacular English," "Spoken Soul," or "Ebonics" (from *ebony* plus *phonics*—literally "black sounds"). William Labov was, in 1969, the first linguist to conduct a formal analysis of the dialect's grammar, showing that, contrary to claims that it is "slang" or "lazy" or "sloppy" speech, Black English (as I will call it) is as complex, consistent, rule-based, and richly expressive as any other dialect (if not more so).[39] Labov offered this analysis in the face of "expert" opinion from school psychologists of the early 1960s that Black English is a "non-logical mode of expression"—that is, *not a language*—and evidence of a genetic deficiency, *a cognitive deficit*, the supposed proof for which were the low reading scores of school kids in the inner city.[40]

Labov showed that this was grotesque racist nonsense, and that the struggles of elementary school African Americans, reared in the Black English typically spoken in the intimacy of the home, reflected nothing more mysterious than the children's lack of familiarity with the Standard English spoken by their teachers. This might suggest a degree of racial segregation bordering on South African apartheid—and, sure enough, Labov showed that the *majority* of inner-city Black children reached fourth grade without ever having had a face-to-face conversation with a white person, and thus *never having conversed in Standard English.* That they had been bombarded, since birth, by the electronic sounds of Standard English blaring from TVs and radios was irrelevant: as we noted earlier, people cannot passively acquire languages, accents, or dialects from an electronic, disembodied voice. Voices are shaped in active interplay with other voices during a critical window in childhood development.

In 1996, the Oakland, California, school board took up a suggestion by Labov and others that teachers speak to children in Black English (or, as it was now being called, Ebonics) as a way to ease the kids into proficiency with Standard English. This was met by a cyclone of outrage from both white and Black commentators who misapprehended it as a proposal to teach kids a nonstandard dialect, *in place of Standard English.* Stanford University sociolinguist John Russell Rickford documented in his book *Spoken Soul* (2000) the widespread ridicule and contempt aimed at the Oakland proposal—and at Black English itself, a mode of speech that for a huge number of African Americans is vital to their sense of personal and group identity, culture, family, and history. Nevertheless, media outlets derided it as "mumbo jumbo," "mutant English," "broken English," "fractured English," "slanguage," and "ghettoese."[41] The right-wing political TV personality Tucker Carlson, then of CNN, now of Fox News, dismissed it as "a language where nobody knows how to conjugate the verbs."[42]

There is debate over the origins of Black English. The "Africanists" insist that it derives from the West African languages spoken by the original slaves wrested from their homeland, a diverse group of speakers who, plunged into the unfamiliar world and language of their new circumstance, were obliged to improvise a pidgin (a simplified form of a language) that bridged the gap between their respective African tongues and English; this then became a creole (a more systematic, grammatically governed speech) spoken by *their* children, and then a full vernacular within a generation or two. Certain grammatical features (like using "be" for permanent situations, as in "She be working") *do* seem to be unique to African tongues. But the "Anglicists" argue that every feature of grammar and accent can be traced to the seventeenth-century English spoken by the white indentured servants (imported from Britain's non-RP-speaking underclass) whom the slaves worked alongside on plantations. Still others suggest that the dialect derives from a merging of the two strains[43] (for which there seems to be compelling evidence).

Regardless of its origins, Black English's continued existence, a century and a half after the emancipation of the slaves, and its growing divergence from Standard English, is testimony, Labov says, to the institutional racism that has characterized the Black experience. And not only in the South. In search of better jobs and better treatment, some six million African Americans fled north in the Great Migration (1916–1970)—only to face discrimination almost as revolting as that which they experienced on the plantations: obliged to live in outlying, decrepit, or otherwise substandard neighborhoods, isolated by white flight to the suburbs, pushed into underperforming schools, denied well-paid (or any) jobs, stopped and frisked, gunned down by police, linguistically profiled, and imprisoned en masse (in 2020, despite a sharp decline in incarceration rates for African Americans, they are still locked up at a rate more than five times that of whites).[44] According to Labov, Black English's continued divergence from the English spoken by the majority white population is a result of the same dynamic that he first documented in the fishermen on Martha's Vineyard—an unconscious rejection of an oppressive, exploiting "other." Or perhaps not so unconscious. Urban anthropologists in the 1980s discovered among Black teens in Washington, D.C., a clear hostility to "acting white"—with speaking Standard English (including "white pronunciation patterns or accents") topping the list of behaviors to be avoided at all costs.[45] And who could blame them?

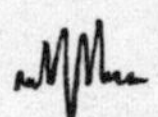

Labov has stated that any public discussion of Black English "generates emotional waves whose violence cannot be overestimated."[46] As a Caucasian (from Canada, no less), I certainly feel considerable trepidation wading into such troubled waters. And I haven't even strayed into the deepest water of all: the question of whether African American voices actually sound different from white voices, apart from the accent and grammar elements I've been discussing—whether some other sonic signature distinguishes them. According to John McWhorter, an author and professor of linguistics at Columbia University, this is a subject that *no one* relishes discussing. "African Americans take it as an insult," he told me, "because Black voices are so often associated with stupidity; while white people avoid it for fear of sounding racist. There is the whole idea that you are not to stereotype, you're supposed to see people as individuals."[47]

But McWhorter, who is Black himself, has made a career of speaking up on matters that others sidestep, and he has identified nuanced aspects of pronunciation (subtle refinements of the Black voice described by Labov, Rickford, Baugh, and others) that, he says, make virtually all African Americans sound distinct from other English-speaking Americans. He calls it the "blaccent," and he says that every American (Black or white) is sensitive to it, whether they admit it or not. "If you're sitting on the subway, looking into your book, and you hear someone's voice, you can know *instantly* whether it is a Black person," McWhorter told me.

"You are almost never wrong. To be an American is to have that ear. And I thought, it's time for somebody to lay out why that is."

He did so in his 2017 book, *Talking Back, Talking Black*, where he says that African Americans have a subtly different way of shaping certain vowels that is unique to a lineage of people whose speech sounds have come down, ultimately, from the original slaves, transferred, generation after generation, by Motherese, and thus etched, at birth, into the basal ganglia and myelinated into the motor circuitry. "It's learned in the cradle!" McWhorter told me. "I find that scientifically *fascinating*—something that happens to people so early and is so deeply imprinted." So deeply, he says, that these tiny vowel "frills" (as he calls them) are manifest even in the speech of African Americans who, in every other way, speak in the most generic, broadcast-ready GA accent. "Even among highly Standard-sounding Black announcers and newscasters," he writes, "*extent* sounds a bit like *extint*, *sense* will sound a bit like *since*, and *attention* will sound rather like *attintion*."[48] To catch the distinctive African American short *o* and *a* vowels, McWhorter invites his reader to

> imagine how Chris Rock would say "Got that?" The difference between how he would say that and Rachel Maddow would nails the nature of these vowels in a light blaccent: as subtle as it is, it is part of what reads "black" in the back of an American mind.[49]

Elsewhere in the book, McWhorter addresses the possibly even touchier question of whether there is a quality to Black voices that is *not* a matter of articulation, grammar, accent, "blaccent," or vocabulary but is, instead, found in some ineffable difference in vocal acoustics, the quality, or timbre, of the sound itself. The question arises if only because decades of listening to James Earl Jones's and Morgan Freeman's resonant voice-of-God baritones, or Oprah Winfrey's soothingly low-pitched voice, has inculcated into listeners an unconscious association between someone's being African American and speaking in a rich, slightly more bass-heavy vocal tone. If such a thing does distinguish some Black voices from white ones, McWhorter says, it emerges not from any anatomical, physiological, or biological difference, but purely from *cultural* factors, a learned tendency picked up in earliest childhood, from listening to adult voices, and mimicking them.

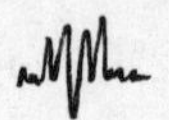

For all the pussyfooting and apologizing that academics and authors (like myself) display when broaching the question of whether Black and white people in America sound different from each other, Black *comedians* have been happily capitalizing on the distinctions for decades.[50] In the 1978 movie *Live in Concert*, Richard Pryor famously code-switches from his characteristic Black English stage voice—a magnificently supple and varied instrument—into a pitch-perfect "white guy voice." The hilarity, and accuracy, of the bit (which hinges on a white man who nervously gives up his place in line to a group of Black men) depends on Pryor's pinched nasality, near-falsetto monotony, rhythmic rigidity, and utter lack of anything resembling the state of being that African Americans taught the world to describe as "cool." ("*Go ahead, sure, cut in*"—and in panicked aside to his white friend: "*What do you want:* trouble? *There's a* whole bunch *of 'em*!"). Since then, many African American comedians have incorporated a "white voice" bit into their routines, including Whoopi Goldberg, Martin Lawrence, Dave Chappelle, and Eddie Murphy.

In the summer of 2018, *two* different Hollywood movies by Black directors took for granted that white and Black people in America sound different from each other: Spike Lee's *BlacKkKlansman* and Boots Riley's *Sorry to Bother You.* Both featured Black protagonists mistaken, over the phone, for white men. In Lee's movie, undercover cop Ron Stallworth uses his "white voice" to converse with Ku Klux Klan grand wizard David Duke (who never suspects he's talking to a Black man); in Riley's movie, a hapless telemarketer, played by Lakeith Stanfield, can't make a sale when speaking in his normal voice. His coworker, played by Danny Glover, offers some advice: "Use your white voice." Stanfield's character snorts and says, dismissively, "You mean like *this*?"—and pulls out a version of Pryor's "white guy" routine. Glover shakes his head. "I'm not talking about sounding all *nasal*," he says. "It's like: sounding like you don't have a care. Got your bills paid, you're happy about your future."

Here, director-writer Boots Riley digs for something deeper than the sardonic laugh of Spike Lee's movie, or Pryor et al.'s brilliant impressions. Riley (as committed a socialist as George Bernard Shaw ever dreamed of being—indeed, he confesses to being an actual *communist*) shows that the distinctions between white and Black voices are, at bottom, distinctions of *class* and economics—the same thing that Shaw was writing about in *Pygmalion* and that Gatsby was talking about when he said Daisy's voice was "full of money." F. Scott Fitzgerald himself could not have put it better than Riley does when he has Glover continue: "Put some real *breath* in there. Breezy. Like: '*I don't really need this money.*' You've never been fired—only *laid off*." And then Riley delivers the coup de grâce, that takes the speech beyond any consideration of "race"—which, after all, is not a biological or genetic reality but a purely cultural construct. We are all *Homo sapiens*, all brothers and sisters under the skin, all in possession of this remarkable endowment of speech that (as Edward Sapir pointed out in the 1920s) is equally complex and expressive in every tongue, and differs only in the surface details dictated by history, culture, and what you learn in the cradle.

"It's not really a *white* voice," Glover's character says. "It's what they *wished they sounded like*; what they think they're *supposed* to sound like."

In 2012, Labov published a deceptively slim volume called *Dialect Diversity in America* that managed, in under two hundred pages, to compress everything he had learned over the previous five decades about American accents and the social, economic, political, and racial forces that shape them. The book included findings from a remarkable project he had undertaken at the beginning of the 1990s when he scaled up his study of progressively larger and larger "islanded" populations (first Vineyard fishermen, then Manhattanites, then the city dwellers of the Inland North, then African Americans) to embrace an "island" of speakers that included all of the United States and Canada—a fifteen-year project that resulted in the monumental *Atlas of North American English* (2006).[51]

The chief surprise to emerge from the project is that, contrary to widespread belief, regional accents in America are *not* vanishing and voices are *not* converging on a single, idealized GA accent inculcated by electronic mass media. Quite the opposite. American voices are *diverging*. The Northern Cities Shift is not a unique development, but just one of a set of "vigorous new sound changes" in progress across the entire country. (In *Dialect Diversity,* he would write: "The dialects of Chicago, Philadelphia, Pittsburgh, and Los Angeles are now more different from each other than they were 50 or 100 years ago.")[52] The *Atlas* also drew on findings by Labov's student Penelope Eckert about the voice that has, in the last few decades, become the stereotype for laid-back California. Eckert traced its origins to Moon Zappa's 1982 pop hit "Valley Girl," in which she (in a rap-like spoken-word section) comically exaggerated the drawled, fronted vowels ("Like, *tewtilly*"), vocal fry, and upspeak of girls from affluent sections of the San Fernando Valley. These mannerisms,

Labov says, became a kind of contagion, like Yeager's cockpit drawl and the Kardashian fry (of which the song might have been the initial germ); the voice spread from the Valley (with later help from the 1995 box office smash *Clueless*, about rich kids in Beverly Hills) and is today a vocal signal of prestige for young women in wealthy enclaves across the country. Although it did not begin as an accent, in the strict sense of pronunciations that become hardwired in the cradle, the Valley Girl voice is what linguists call a "stylistic innovation." By now, however, it is almost certainly being inculcated into newborns by their "chill-talking" parents and can legitimately be called a new American accent.

That the divergences in American voices are shaped by sociolinguistic forces—clashes of ideology, culture, class, politics, and race—is a fact that hovers uneasily over the *Atlas of North American English*, whose pages are replete with maps that show, superimposed on the fifty United States, a spaghetti-tangle of lines (linguists call them "isoglosses") that define the borders between accent regions. Given current sociopolitical realities, it is hard to see these isoglosses as anything but fault lines traced on a dangerously unstable seismographic map. Labov acknowledges as much when, in *Dialect Diversity*, after describing the incredible fact that vowel-shifting people in the Inland North are misunderstood by *each other*, he writes that the divergence of voices, in all societies, "runs counter to the primary function of language as it has developed in the human species: the capacity to communicate information about states of affairs across distant times and places. We are not better off because we do not understand each other."[53]

Labov here hints at the prime paradox of the human voice: at the same time as its specialization for language has united our species and given us dominion over every other living thing by bringing us together into cooperative groups, it has, at the same time, "pushed us apart"—and continues to do so. In the *Origin of Species*, Darwin used the way languages change over time as a metaphor for natural selection. Like animal species, he said, languages are evolving toward *greater efficiency*, which would mean enhanced understanding between all human speakers. This is one of the few times Darwin got it wrong—because this is not at all what happens with language. The exact opposite. It is precisely through the incremental, natural-selection-like accumulation of tiny, random changes in articulation—McWhorter's "frills" on the vowels, the migration of an *o* or an *a* backward or forward in the oral cavity, that slight shift in a tongue target from teeth to gum ridge, a change in cadence or intonation like Valley Girl upspeak—that the Babel of mutually incomprehensible tongues, across human populations, came about. It is an amazing philological fact that languages as different from one another as Russian, Hindi, Greek, Albanian, French, German, English, and Icelandic all descended from a single group of Proto-Indo-European dialects that ultimately had to have developed from some single, original, and now unrecoverable ur-tongue. This means that our greatest endowment as a species—the spoken language that has allowed us to converse, coexist, compromise, and cooperate in a manner unique in the animal kingdom—is also what renders us dangerously unintelligible to one another. And pushes us apart.

Which is where political leaders come in: those people who shape our collective future, and whose task it is to unite people of widely differing views and values, colors and creeds, attitudes and accents; who instill within us—primarily through the galvanizing force of *their* voices raised in oratory—a sense of common purpose, of our shared humanity and shared challenges and goals, so that we might bridge our differences of class, race, education, religion, politics, gender identity, sexual orientation and even nationality, to forge, for the betterment of the species as a whole, a bright and blessed future for all humankind!

That's the idea, anyway.

SEVEN

THE VOICE OF LEADERSHIP AND PERSUASION

The idea of democracy, "rule by the people," was dreamed up, around 500 BC, by the ancient Greeks, who, even as they extolled it as the best way for human populations to govern themselves, also warned of a danger inherent in choosing leaders by popular vote. This opened society to the risk of charismatic, self-dealing, narcissistic charlatans who attract voters not through an appeal to reason (as the Greek model of *dēmokratía* demanded), but through a whipping of pure emotion. The Greeks called them *demagogues*, and political scientists have since shown how these opportunistic threats to civilization use lies, distortions, and fear-mongering to awaken our species' worst instincts, including class and racial prejudices, xenophobic hatred of the other, and myriad other societal divisions to win votes. Speaking in incendiary tones of rage, recrimination, and vengeance, the demagogue is swept to power on a wave of angry-mob "populist" sentiment.

One of the demagogue's chief means for attaining power is through misappropriation of the skills of vocal persuasion that the ancient Greeks and Romans considered essential to democracy: oratory and rhetoric. *Oratory* is the formal, elevated style of public address used by politicians in speeches and debates (as well as preachers speaking to congregations and lawyers addressing juries). *Rhetoric* is the orator's eloquent and inventive use of language—the vivid similes and metaphors, the sensitivity to the sounds of words, the skillful compression, the apt word choice, the sonically similar (or contrasting) grammatical structures that bolster logical arguments, and which hold listeners' attention and drive ideas home. In Greece and Rome, fine oratory and rhetoric were, says one leading classical scholar, "the lifeblood of ancient politics, law and administration, a shared discourse that enabled communication across boundaries of ethnicity, status and ideology."[1]

Some ancient philosophers (including Plato) worried that oratory and rhetoric are a form of deception—performance techniques to hoodwink listeners. But Quintilian (circa AD 35–100) insisted that they are not surface tricks of voice and speech—auditory masks—that conceal the speaker's true character and intent. Properly practiced, they reveal the truth of who a person is: crystallizing, in the voice signal, the speaker's personality, intellect, values, and intentions, so that the person's arguments may be scrutinized, for accurate evaluation, by voters, juries, or religious congregations. Like the skills of economy and simplicity that Strunk and White endorse in their classic writing manual, *The Elements of Style*, oratory and rhetoric are "the Self escaping into the open." To be more accurate, Strunk and White said that *all* writing ultimately exposes the true self, even bad writing. However, good, vivid, forceful, well-reasoned writing is an index to the essential sanity and moral character of the writer. According to the ancient philosophers, the same is true of fine rhetoric and oratory.

Cicero expressed this with special urgency in his treatise *On Oratory* in 55 BC, when the Roman Empire was tilting toward civil war. Because demagogues thrive in periods of political and social instability—when a demoralized, angry, or fearful populace is most vulnerable to atavistic calls for vengeance and violence—Cicero deemed it an opportune moment to remind his fellow Romans of the qualities that make up the ideal orator, whom he describes as a supremely moral person, a cultured citizen versed in disciplines ranging from "language and literature to politics, psychology, law, history, aesthetics"—even "child development."[2] Otherwise, he said, "the flow of words is empty and ridiculous." According to Cicero, the skillful orator achieves three aims: *docere, delectare, et movere*—that is: he (and in ancient Rome it was invariably "he") *proves* his thesis to his audience, he *delights* his audience, and he moves it *emotionally*—although not through the demagogue's resort to yelling and hate speech. Emotion is instead engendered by an artfully poetic and euphonious language that is part of a larger *rational* argument that wins listeners to a just cause.

Cicero catalogued many rhetorical devices for persuasive public speaking; some aimed at supplying the liveliness and beauty that he believed necessary for holding an audience's attention by pleasing the ear. These included *alliteration*, the artful repetition of consonant sounds (like the *s*'s in the opening of Shakespeare's sonnet 30: "When to the sessions of sweet silent thought / I summon up remembrance of things past . . ."); *assonance*, the deliberate repetition of vowels (like the *o*'s that come later in that sonnet, "And heavily from woe to woe tell o'er/ The sad account of fore-bemoaned moan"); but also grammatical devices like *asyndeton*, the piling up of clauses with no conjunctions between them, which lends vigor and intensity to speech, as in Caesar's boast *veni, vidi, vici* ("I came, I saw, I conquered"); or a device like *zeugma*, the witty use of a single word to modify two others, as in, "I broke the mirror and my promise" or "She played the piano and the fool." Cicero also included a host of grammatical constructions, like *antithesis*, in which opposed ideas are expressed in identical form to make a statement powerful and memorable (like Neil Armstrong's words as the first human to set foot on the moon: "That's one small step for a man, one giant leap for mankind"); and various rhythmic variations, each with its own term and definition, that strengthen oral argument by making it moving, memorable, pleasurable—and thus persuasive.

As a writer on rhetoric, Cicero not only shaped how people in Western society, to this day, express their ideas in *speech*, but also (as the examples from Shakespeare suggest) in writing—a distinction that people in ancient Rome were less likely to make than we are. Today "sounding out" text is what first graders do, and the expression "moves his lips when he reads" means "he's stupid." But in Cicero's time, reading invariably meant reading *aloud*, as Ben Yagoda documents in his book *The Sound on the Page* (2004). In the ancient world, Yagoda writes, "a written text was like a play script, only in performance did the words come to life." Indeed, silent reading was so rare that, even by the fourth century AD, Augustine, the early Christian theologian, was amazed at the sight of the bishop of Milan reading a book: "his eyes scanned the page and his heart sought out the meaning, but his voice was silent and his tongue was still . . . he never read aloud."[3]

For almost a millennium after that—until the early 1300s—reading still meant speaking a text aloud; that is until Dante, Petrarch, and Chaucer began writing in the vernacular of their native tongues (instead of Latin). This brought written poetry and prose closer to everyday speech and led to greater private (which is to say, silent) imbibing of the written word; Gutenberg's invention of the printing press, in 1440, and the mass production of written material solidified the silent consumption of texts. But Cicero's rhetorical rules for effective and persuasive *speaking* had been so thoroughly absorbed into writing by then that they informed the work of every writer in virtually every tongue, and still do. Alexander

Pope's poem "Sound and Sense," written in 1711, is *about* how poetry (and indeed prose) succeeds only inasmuch as it evokes the sounds of speech, the particular way that voiced language mirrors the meaning of words. "'Tis not enough no harshness gives offense," Pope writes, "The sound must seem an echo to the sense"—and he ingeniously uses the eight lines that follow to illustrate the point with an array of rhetorical strategies straight out of Cicero: multisyllabic words that make lines that describe speed move rapidly; one-syllable words that describe ponderous movement and make words move slow; consonants like *s* and *z* that whisper and sing for describing sweet sounds contrasted with *x*'s and *f*'s and percussive *t*'s that describe noise, plus a complex interweaving of alliterative and assonant sounds that hold the whole thing together and make it beautiful and memorable. For maximum effect, you should do like Augustine, and speak the lines aloud:

> *Soft is the strain when Zephyr gently blows,*
> *And the smooth stream in smoother numbers flows;*
> *But when loud surges lash the sounding shore,*
> *The hoarse, rough verse should like the torrent roar;*
> *When Ajax strives some rock's vast weight to throw,*
> *The line too labors, and the words move slow;*
> *Not so, when swift Camilla scours the plain,*
> *Flies o'er the unbending corn, and skims along the main.*

Thanks to Cicero and the other ancient writers on rhetoric, the best literature is always aimed at the ear and the speech centers. After all, Shakespeare began life as a *voice artist*—an actor—and wrote his plays purely to be heard, so much so that, as scholar G. Blakemore Evans notes,[4] Shakespeare took no interest, in his lifetime, in his plays being properly printed in authoritative editions for *readers*, allowing them to circulate in quarto editions of vastly varying quality and accuracy, the "bad" quartos reconstructed from memory by literary pirates (with the expected omissions and mistakes); the "good" quartos, although printed for the use of Shakespeare's theater company, often full of errors, typos, dropped lines. The First Folio edition (1623), which included the "authoritative" texts of all Shakespeare's plays, was the initiative not of Shakespeare, but of two actors in his theater company—and not compiled and published until seven years after Shakespeare's death. Clearly, for Shakespeare, the plays that we recognize as the zenith of written language, were composed purely for speaking. Keats was thinking about the sensuous dance of lips and tongue—not black marks on a page—when (in his ode to the mellifluous voice of the nightingale) he described a glass of wine with "beaded bubbles winking at the brim," and even the most opaquely literary of twentieth-century modernists, like Joyce, wrote directly for the voice and ear: the polyglot punning of *Finnegans Wake* can only be deciphered if read aloud. Nabokov, who raised euphony to levels verging on purple prose, begins his most famous novel by evoking an erotic obsession through the sensual pleasure of moving the articulators (especially, of course, the tongue), while relying on the alliteration and assonance that Cicero said were critical to memorable, delightful speech: "The tip of the tongue taking a trip of three steps down the palate to tap, at three, on the teeth. Lo. Lee. Ta."[5]

Cicero's influence on Western speech was not fully felt until the Renaissance when the complete text of *On Oratory* was discovered in 1421 in the town of Lodi, near Milan. In England, his precepts were avidly taken up because British parliamentary debate—the custom of the opposed political parties coming together, in public, to shout, fulminate, jeer at, and heckle one another—put great emphasis on the ability of politicians to organize and deliver arguments effectively. Cicero's impact on public speech in America has been no less profound, although slower in adoption. "George Washington made a retiring and awkward speaker," writes Carolyn Eastman, a professor at Virginia Commonwealth University and a specialist in the cultural history of early America. "Thomas Jefferson could barely speak above a whisper, and the former lawyer John Adams was divisive and abrasive rather than persuasive."[6] But within a generation, American public speech was already starting to improve under the influence of the ancients. When, in 1805, Adams's son, John Quincy Adams, was hired as the country's first Professor of Rhetoric and Oratory at Harvard (twenty years before he became the sixth U.S. president), he used his Harvard inaugural address to emphasize the supreme importance of Cicero and the other ancients to effective public speaking. "In the flourishing periods of Athens and Rome," he said, "eloquence was power."[7]

Even as John Quincy Adams was promoting classical rhetoric and oratory within the ivory tower, ordinary Americans were getting a galvanizing taste of the real thing. In 1808, three years into Adams's appointment at Harvard, James Ogilvie, a Scottish-born elocution teacher in Virginia, noticed the massive popularity of public lectures across America—a form of mass entertainment that drew thousands of paying customers to theaters and outdoor venues to hear speeches, readings, debates. Ogilvie would go on to become the most popular of them all when he quit teaching to become an itinerant orator, traveling the country's lyceum circuit, a hugely popular movement of the early nineteenth century that organized the various forms of "platform culture" into ticketed performances.[8] Ogilvie spoke in a style that, Eastman says, "mixed emotion and reason in a mesmerizing swirl." Newspapers advertised his upcoming appearances, thousands turned out to hear him ("from Georgia to Maine and from Tennessee to Québec"), and a raft of imitators began to mimic his oratorical style (and also his costume, which emphasized his debt to the ancient orators: *a toga*). Ogilvie also spoke at the Capitol, in addresses to President Madison and both houses of Congress—which helped to impress upon America's political class the importance of Ciceronian rules of public address. Eastman says that Ogilvie's influence would soon be seen in the emergence of oratorically gifted House and Senate leaders, among them Henry Clay, Daniel Webster, and eventually, in the early summer of 1858, a candidate for U.S. senator from Illinois, Abraham Lincoln.[9]

Lincoln's public speeches would shape the United States to a greater degree than anyone's before or since. Indeed, it says something about the momentousness of his influence as a speaker that the *least* of his accomplishments as an orator is to have invented a staple of the modern electoral campaign: the political debate.

Beginning in the late eighteenth century, the elocution movement in America had given rise to "debating clubs" where, for the entertainment and edification of audiences, the participants engaged in disputation on the model of the British Parliament, with one side taking the role of the "government," the other the "loyal opposition," and arguing current topics in science, the arts or politics. Lincoln, however, invented the political *campaign* debate when as a little known country lawyer and former one-term congressman, and underdog challenger for the U.S. Senate, he challenged to a vocal showdown the well-known, well-funded, two-term Democratic incumbent, Stephen A. Douglas. After doggedly following Douglas on campaign stops around Illinois (and jumping onstage to address whatever stragglers still remained after Douglas, the "main attraction," had quit the stage), Lincoln finally managed to goad Douglas into direct debate. In a letter of July 1858, he proposed that he and Douglas "divide time, and address the same audiences"—that is, share the campaign stage and block out chunks of time to vocally duke it out. It had never been done before. Douglas, as the well-known front-runner, had everything to lose, Lincoln (a virtual unknown) everything to gain, but Douglas recognized that he might be seen as weak or fearful should he spurn Lincoln's challenge. Besides which, Douglas had every right to assume that he would prevail.

Douglas possessed a big, commanding baritone and used it with confidence and great oratorical flourish. Lincoln's voice, in contrast, was famously reedy and high-pitched, his delivery sometimes halting. The journalist Horace White, who wrote an eyewitness account of the debates, described Lincoln as having a voice "almost as high-pitched as a boatswain's whistle."[10] (Others called it "shrill," "sharp," and even "unpleasant.")[11] Daniel Day-Lewis, who played Lincoln in the 2012 Spielberg film, has spoken of "finding" this voice through researching Lincoln's character, studying his speeches, and considering his acts—thus arriving at a voice that seems not a surface "impression," but instead a deep-dyed aspect of Lincoln's character—a "kind of a fingerprint of the soul," as Day-Lewis put it to Oprah in an interview about the movie.[12] It's impossible to say how close Day-Lewis's "Lincoln" voice is to the real thing: the first recording device did not appear until 1877, twelve years after his assassination. For my part, though, I could never hear, in my "mind's ear," Lincoln's reportedly high, thin voice until Day-Lewis's squeaky barn-door hinge interpretation; now it's the only thing that makes sense to me.

Lincoln's pitch and timbre were not his only seeming vocal handicaps in his face-off with Douglas. Lincoln's off-the-cuff speechmaking left some things to be desired: "his words did not flow in a rushing, unbroken stream like Douglas'," White wrote. "He sometimes stopped for repairs before finishing a sentence. . . . After getting fairly started, and lubricated, as it were, he went on without any noticeable hesitation, but he never had the ease and grace and finish of his adversary."[13] But as the debates proceeded, from late summer through early fall, Lincoln's perceived weaknesses began to look more like strengths: his high, piercing voice penetrated to the back of the ever-growing crowds (while Douglas, perhaps pushing too hard to be heard, suffered an ill-timed attack of laryngitis in the final two debates). Even the pauses and hesitations in Lincoln's speech might have been advantageous, perhaps communicating, paralinguistically, a thoughtfulness and authenticity in contrast with Douglas's more "flowing," facile, and *slick* oratory: as a politician, you can be, perhaps, too eloquent.

Most important to Lincoln's performance in the Douglas debates, however, was the righteousness that fueled his tone. For the topic of all seven debates concerned an issue consuming the Union as it expanded westward: whether the new territories should make slavery legal. Lincoln considered slavery morally repugnant and completely contradictory to the Declaration of Independence and its foundational democratic premise that "all men are created equal." These convictions fired his voice with what White called a "moral superiority and blazing earnestness that came from his heart and went straight to those of his listeners."[14] Douglas, meanwhile, argued that *his* vision of democracy was correct because it gave the issue of slavery to the states to decide, by democratic vote. (This was a cover. Douglas had made his repulsive moral position of white supremacy clear in a speech earlier that summer—a speech that, indeed, helped spur Lincoln's determination to take him on in live debate: "I am free to say to you that in my opinion this government of ours is founded on the white basis. It was made by the white man, for the benefit of the white man, to be administered by white men.")[15]

Although Lincoln won the popular vote, he lost the election to Douglas through the vagaries of districting and vote-counting. However, the debates, which had been witnessed in person by tens of thousands and followed by millions more in extensive newspaper coverage, instantly propelled Lincoln to national prominence. Two years later, in 1860, he published the debates and the book became a bestseller and a major promotional tool for his presidential campaign, launched that same year. When he won, Lincoln said he was "accidentally elected" because of his debates with Douglas.[16] Accident or no, the debates marked a moment when a single speaking voice changed the course of American history, giving rise to a presidency that freed the slaves, prompted the succession of the Southern states that then launched a civil war, and resulted ultimately in the Union victory that prevented the breakup of the nation. Lincoln's remarkable presidency also led ultimately to his assassination—but not before he delivered a speech less than two years before his death in which he mused that "the world will little note, nor long remember what we say here." Today, historians describe this speech of a mere 272 words as "the most eloquent articulation of the democratic vision ever written"[17]—a memorial address for the Civil War dead at Gettysburg.

Lincoln, whose public speeches tended to an American plain-style, began the Gettysburg Address with a rare rhetorical flourish—"Four score and seven years ago"—a choice that E. B. White, in *The Elements of Style*, says was clearly driven, not by literary considerations, but *oratorical* ones, those involving the voice, and dictated by the ear. "The President could have got into his sentence with plain 'Eighty-seven,'" White says, "a saving of two words and less of a strain on the listeners' powers of multiplication."[18] But by "skirting the edge of fanciness," White says, Lincoln "achieved cadence"—and (I would add) gave a fitting sense of momentousness to that span of years since the founding fathers fitted their signatures to the Declaration of Independence and its foundational idea—the idea that fueled Lincoln's entire political career and that Thomas Jefferson had memorialized in the words: "all men are created equal."

Even as Ogilvie, Douglas, and Lincoln were using Ciceronian rules of rhetoric and oratory to shape political speech and debate in America, another strain of address was influencing the "platform culture" of the eighteenth and nineteenth centuries: the sermon. Ancient rules of rhetoric and oratory had long informed preaching in many religions. So closely did Christian sermonizing of the thirteenth century adhere to classical rules of oratory, Cicero's image (as well as that of "Rhetoric") is carved into the facade of Chartres Cathedral.[19] The synagogue sermons and Talmudic disputations of Judaism derive directly from Cicero's rules of oratory and debate, as does the Passover Seder, a ceremonial service at the beginning of the high holidays whose "script" follows the precise structure Cicero laid out for legal addresses, beginning with an *Exordium* (introduction), followed by *Narration* (in which facts are enumerated), *Partition* (a breakdown of proofs to come), *Confirmation* (summary of evidence), and a *Conclusion* (that recaps the argument).[20] Preaching in Islam also uses Greek and Roman rhetorical models (Aristotle's writing on rhetoric and oratory were translated into Arabic in the eighth century), but Islam also views classical models of preaching with some suspicion, because philosophy is considered incompatible with religion, and the Prophet Muhammad's eloquence was "unsurpassable."[21] Nevertheless, the sermons delivered by imams employ many of the devices described by Aristotle, Cicero, and Quintilian for providing pleasure and persuasion in preaching, including the use of alliteration, antithesis, and hyperbole.[22]

The Roman Catholic oratory that so terrifies Stephen Dedalus in James Joyce's *A Portrait of the Artist as a Young Man* is a model of Ciceronian rhetoric. Father Arnall's minutely detailed depiction of hell (the stink, putrefaction, burning pain, piled corpses, devouring worms, and unending physical and mental torture) are drawn directly from Dante's *Inferno*, which was written in the fourteenth century and based on classic rhetoric. But a still more visceral, emotional style of Christian oratory had also arisen, in the early 1730s in Britain, among Protestant evangelical preachers whose sermons stressed the importance of spiritual rebirth and the speaker's personal salvation. Known as "fire-and-brimstone" or "hellfire" sermons, they arose in direct reaction to the Enlightenment and the drift away from the church toward rationality and science. Imported to the North American colonies by British missionaries in the early eighteenth century, when church membership had drastically declined, this exceedingly emotional preaching style spurred a wave of revivalism, a passionate Puritan religiosity known as the First Great Awakening. The most famous of these revivalist sermons, "Sinners in the Hands of an Angry God," was written and delivered by pastor and theologian Jonathan Edwards, and subsequently published in 1741. Like Dante in the *Inferno* (and Father Arnall in *A Portrait*), Edwards depicts Hell as a *real place* where sinners will end up—the difference being Edwards's characterization of hell's proximity and immanence, and of God, not as loving Father, but as a vengeful entity almost eager to consign humanity to hell:

> The God that holds you over the pit of hell, much in the same way as one holds a spider, or some loathsome insect over the fire, abhors you, and is dreadfully provoked; his wrath towards you burns like fire; he looks upon you as worthy of nothing else but to be cast into the fire; he is of purer eyes than to bear to have you in his sight; you are ten thousand times more abominable in his eyes than the most hateful venomous serpent is in ours.[23]

Surprisingly, Edwards's delivery of this sermon (first to his congregation in Northampton, Massachusetts, and later to a parish in

Connecticut) was oratorically subdued: he read from his prepared text in a near monotone, eyes glued to the page.[24] This was unusual, since fiery vocal pyrotechnics, dramatic physical gestures, and blistering extemporaneous deviations from written texts became hallmarks of the evangelical preaching style, especially during a later wave, known as the Second Great Awakening, which occurred at the dawn of the nineteenth century in America. The fire-and-brimstone Baptist and Methodist preachers of the Second Great Awakening were especially prized for their ability to go off-text, in rhetorical and oratorical flights grounded in the most minute knowledge of scripture. Their highly emotional delivery—shouting, quavering, whispering, sobbing—were elements that demagogic politicians would borrow.

YouTube abounds in examples of today's evangelical preaching style, which, all evidence suggests, differs little from that of the early nineteenth century. In a video posted on January 11, 2011,[25] Pastor Charles Lawson of Temple Baptist Church in Knoxville, Tennessee, can be seen and heard offering a softly drawled greeting to his congregation, before introducing the topic of that day's sermon: Christ's Sermon on the Mount—although *not*, Pastor Lawson stresses, the part beloved of "liberals," about *turning the other cheek*. Instead, the pastor says, he will speak on Jesus's little-noted mention, near the end of the Sermon on the Mount, of "hell-fire." Pastor Lawson then launches into a sustained, twenty-five-minute harangue on the absolute, real, true, undeniable existence of *hell*, and the certitude that everyone will end up there unless they repent. What the sermon lacks in Father Arnall's exquisite descriptiveness in *A Portrait of the Artist*, it makes up for in oratorical passion. By the ten-minute mark, Pastor Lawson has worked himself into an arm-flinging tirade, bellowing in a timber-rattling voice that swells in volume with each new utterance of the word *hell*—or *Hay-Yull*, in the pastor's Southern pronunciation. Some Ciceronian rhetorical moves *are* featured—especially the rhythmic use of repetition to build an emotional crescendo—but it is secondary to the sheer volume and fire of the pastor's voice: "There is no sal*va*tion in Hay-Yull! There is no *sa*vior in Hay-Yull! There is no *Bi*ble in Hay-ullll! There is no blood in Hayyy-yull! There is no for*give*ness in *Hayyy-yulll*!" Then, in a rattling bellow identical to the throat-ripping vocals of a death metal singer, he shrieks: "*Whatever GOES to Hayy-yulll STAYS in Hayyyy-Yull!*" Easy to mock and parody, the effect is genuinely frightening and leaves little doubt about how the evangelical Baptist and Methodist preachers of the First and Second Great Awakenings made their churches the most popular across the Southern United States. They remain so to this day.

Enslaved Blacks were among those that the traveling Baptist preachers of the Second Great Awakening especially targeted for conversion, promising leadership roles in the church (roles denied to them in Anglican and Episcopal denominations). Baptist slave preachers found particular salience (and solace) in Bible stories of survival under oppression and deliverance from servitude (like the liberation of the Israelites from slavery in Egypt). Indeed, the slave-turned-preacher Nat Turner was sufficiently inspired by such stories that he mounted in 1831 an armed rebellion of enslaved and free Blacks in Virginia. He was captured and executed. This, and other slave rebellions, ultimately led to legislation outlawing Black congregations without a white minister present. Underground all-Black churches duly sprang up, in which a uniquely African American style of preaching and worship flourished.[26]

Mixing highly emotional evangelical preaching with African ideas and rhythms, Black churches created a tradition of lively shouting, singing, clapping, and call-and-response interaction with the pulpit that remain a staple of Black churches today. Geneva Smitherman, an English professor at Michigan State University, documents in her book *Talkin and Testifyin* (1997) the highly specific style of vocal delivery typical of Black evangelical preachers—a melodic, rhythmic style in which "the voice is employed like a musical instrument with improvisations, riffs, and all kinds of playing between the notes."[27] Smitherman calls this "tonal semantics" and says that it has influenced Black speech well beyond the pulpit. The voice is heard "in the speech-music of James Brown and Aretha Franklin, in the preaching-lecturing of Martin Luther King, Jr., and Jesse Jackson, in the political raps of Stokely Carmichael and Malcolm X, in the comedy routines of Flip Wilson and Richard Pryor."[28] And it would be heard, as we will see, in the oratory of the man who would become the first Black president of the United States.

Political speech raises the stakes on rhetoric and oratory precisely because its tools are put in the service, not of "mere" delight (as in literature), or worship (as in religion), but of moving people to decisive action and sacrifice. Few are the instances when a *single* act of political oratory has changed the course of human history, but Prime Minister Winston Churchill's address, on June 4, 1940, to the British House of Commons was arguably such a speech. Coming in the wake of Britain's retreat at Dunkirk and with Nazi forces poised to invade England, the address was expressly designed not only to steel Britons' resolve and to warn Hitler of their fighting spirit but also to signal to a recalcitrant United States that it was time to enter the war and help beat back Germany's plans for world domination. In short, an important address—one in which the fate of the free world hung in the balance. In the 2017 movie *Darkest Hour*, Churchill is shown preparing to compose the speech by pulling from the bookshelf his translation of Cicero's *On Oratory*. Whether that is a Hollywood embellishment or not, Churchill clearly knew that, in the coming battle between democracy and dictatorship, he had to summon the greatest possible eloquence, and he did so in a speech replete with ancient rhetorical devices, particularly in the peroration—or *peroratorio*, Cicero's term for the final rousing appeal to listeners' reason and emotion.

There, Churchill used several rhetorical devices from *On Oratory*, including *anaphora* (deliberate repetition of words), *geminatio* (repetition of particular grammatical structures), and lots of artfully deployed assonance and alliteration. "We shall not flag or fail," he said as he launched into the famous peroration. Though the original speech to the House of Commons was not recorded, Churchill reenacted it, for posterity, in a 1949 home recording in which he delivers the address in a voice of surprisingly quiet resignation, rather than the hortatory declamatory style the words might imply. Whether he spoke in this manner during his address in the House is not known but, on the recording, it is an oratorical choice that, paradoxically, serves to heighten the words' power, underlining a deadly earnestness and resolve. "We shall go on to the end. We shall fight in France. We shall fight on the seas and oceans. We shall fight with growing confidence and growing strength in the *air* [the only word on which his voice stabs upward in pitch and volume, before returning to its steady, resigned, implacable tread]. We shall defend our island, whatever the cost may be. We shall fight on the beaches. We shall fight on the landing grounds. We shall fight in the fields and in the streets. We shall fight in the hills. We shall *never* surrender . . ." According to historian Jill Lepore, in her book *These Truths*,[29] Churchill's speech was broadcast live by radio stations "across America," and it seems to have had the desired effect on Roosevelt's thinking. Six days later, the president, delivering the commencement address at the University of Virginia, spoke passionately about the horrors of a nation under foreign dictatorship, which he described as "the nightmare of a people lodged in prison, handcuffed, hungry, and fed through the bars from day to day by the contemptuous, unpitying masters of other continents."[30] Three months later, he initiated the country's first peacetime draft.

Roosevelt was as effective as Churchill in the use of his voice to rouse the nation he led but his power as a public speaker derived, for the most part, not from Churchill's rolling, orotund, rhetorically rich delivery, but from a speech more reminiscent of Lincoln's American plain-style. Roosevelt first marshalled this, as an inspirational tonic for the nation, eight days after his inauguration in March 1933. With the country sunk in the depths of the Great Depression, Roosevelt had the revolutionary idea of using the new medium of radio to directly address the country. This would initiate a series of addresses that became known as "fireside chats," which he used throughout his presidency to quell panic and stiffen spines during crises that included the near-collapse of the banking system, record unemployment, the surprise attack on Pearl Harbor, and the war against Hitler and Nazism. A single reference to the "phantom of fear" was his only alliterative sally in his debut fireside chat on the banking crisis, and the famous phrase "a date which will live in infamy" (to describe the attack on Pearl Harbor in a speech in Congress) was about as fancy as he ever got—which doesn't mean he was without rhetorical mastery. It is not easy to explain banking and finance to laypeople, but FDR did so with a clarity and precision that allayed public panic over his decision to close the banks and that imbued depositors with sufficient confidence in the financial system to prevent a run on the banks when he re-opened them. The phrase "it is safer to keep your money in a reopened bank than under the mattress" might not sound like deathless rhetoric but such straight talk was key to the remarkable success of those radio addresses. For the first time in history, a world leader put himself on a conversational par with ordinary people, speaking ordinary language—albeit in a steely tone of unshakable confidence that derived, no doubt, from FDR's patrician background, his "ruling-class" upbringing. But even his aristocratically dropped *r*'s ("phantom of *feah*") and lock-jawed pronunciation of long *ee* did not alienate the masses; it was as if his willingness to speak in his *actual* voice, without any patronizing (or inadequate) code-switching, contributed to the sense of directness and honesty he conveyed. Thousands of letters poured into the White House after his debut fireside chat, including one from a listener who wrote, "It almost seemed the other night, sitting in my easy chair in the library, that you were across the room from me."[31]

This intimacy of connection with the American people was also crucial to FDR's garnering the public support that allowed him, in the face of unimaginably fierce political and corporate resistance, to initiate the most comprehensive economic reforms since America's founding—a redistribution of money from the hands of the few to the many (through the advent of Social Security, bank regulations, tax reform, and other measures) that brought the true ideals of democracy to America, by introducing greater economic opportunity for every citizen, in the form of the New Deal—a deal FDR sold directly to everyday Americans through the radio with his own voice.

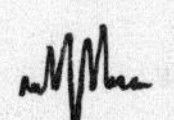

Classic rules of rhetoric and oratory continued to play a decisive role in postwar geopolitics, especially in a series of dramatic showdowns between democracy and totalitarianism. In June 1963, at the height of the Cold War, President John F. Kennedy delivered, in West Berlin, one of his most famous addresses, second only perhaps to his inaugural speech when he invited Americans to "ask not what your country can do for you—ask what you can do for your country"—a nice example of Ciceronian parallelism of sentence construction that made the phrase not only as forceful as possible, but unforgettable.

Kennedy's West Berlin speech came just eight months after the Cuban Missile Crisis, when tensions between the United States and the Soviet Union had brought the two superpowers to the brink of nuclear war.[32] Delivered in the shadow of the newly constructed Berlin Wall, which divided capitalist West Berlin from communist East Berlin, the speech was meant to defuse tensions along a border that threatened to be the next flashpoint for nuclear Armageddon. But Kennedy was horrified by his first in-person look at the looming concrete barrier that the communists had built to keep their people from fleeing dictatorship, and he decided that, with the eyes and ears of the world upon him, he would seize the opportunity to express his genuine outrage at Soviet authoritarianism. He scrapped the anodyne address prepared by his speechwriters and slashed down, in his own handwriting, an idea that derived from his knowledge of classical history. "Two thousand years ago," he scribbled, "the proudest boast was 'Civis Romanus sum' "—*I am a Roman*—"Today, in the world of freedom, the proudest boast is *Ich bin ein Berliner!*" The power and passion with which Kennedy declaimed those words from the stage in front of West Berlin's city hall made the speech justifiably famous as a denunciation of dictatorship, but it is a longer passage of pure Ciceronian oratory that distinguishes this as one of the great speeches of history, made all the greater by the fire of Kennedy's delivery, his Boston vowels elongated so that they carry like musical cadences, his voice tuned to a steely timbre that left no doubt of his genuine disgust at the spectacle of misery he had glimpsed when peering over the top of the wall into a grim and barren Soviet-occupied East Berlin, and a populace imprisoned:

> There are many people in the world who really don't understand—or *say* they don't—what is the great issue between the free world and the communist world. *Let them come to Berlin.* There are some who say that communism is the wave of the future. *Let them come to Berlin!* And there are some who say, in Europe and elsewhere, we can work with the Communists. *Let them come to Berlin!* And there are even a few who say that it's true that communism is an evil system, but it permits us to make economic progress. *Lasst sie nach Berlin kommen*—let *them* come to Berlin!

A historical, and oratorical, bookend to Kennedy's *Ich bin ein Berliner* speech came with an address, also delivered in West Berlin, by President Ronald Reagan, on June 12, 1987, when he aimed at Soviet leader Mikhail Gorbachev a Ciceronian series of escalating admonitions ("If you seek peace . . . if you seek prosperity . . . if you seek liberalization . . .") culminating in the shouted line: "Mr. Gorbachev, *tear* down this *wall*!" Some historians argue that the speech had little impact at the time and was only retroactively enshrined as a moment of epoch-making oratory two years later, when the Wall actually came down (and not because Reagan demanded it). Whatever the case, there is a larger argument to be made that both Kennedy's and Reagan's Berlin speeches are, for all their adherence to Cicero's rules, a sign of the *decline* of classic oratory, given that both speeches are best known for one quotable four-word message (*ich bin ein Berliner*; *tear down this wall*): sound bites; short pithy phrases easily digested by a public not only disinclined to listen to long, involved, rhetorically complex speeches, but who are incapable of it.

This argument hinges on the notion that our species' cognitive apparatus has been refashioned by electronic mass media, fast-edited movies, and TV shows—and now clickbait YouTube videos and TikToks—that have reduced our attention spans to the point where it is impossible for us to follow the spoken arguments in, for instance, Kennedy's *Berliner* speech—to say nothing of the Lincoln-Douglas debates, which were a minimum of *three hours* long each: one hour for an "opening address" by a candidate, an hour and a half for the opponent's reply, and then a half hour for "rejoinders." And people were *riveted*!

Compare this to our televised campaign debates where presidential candidates "discuss" the most momentous issues of the day in rapid-fire *one- or two-minute* verbal sprints. This might suggest that we have indeed suffered a species-wide cognitive decline that would make it impossible for us to attend to Lincoln's and Douglas's voices delineating complex ideas and arguments over several hours of granular verbal disputation. Except that it is not true. We have suffered no such cognitive deterioration and no related decline in our fascination with, and interest in, listening to the human voice, singly or in conversation, as it articulates exceedingly complex ideas over hours of discussion. What *has* happened is that the exigencies of commercial television broadcasting, which demands regular pauses for a word from our sponsor, and which has evolved a Scheherazade-style of breathless *this-just-in* narrative intent on keeping viewers from changing the channel, has whittled down the time that politicians are allowed to speak, for fear (once again) that a bored or restive viewer will reach for the remote. Recent events have shown that we not only *can* listen to complex spoken argument for hours at a time, we love to do so.

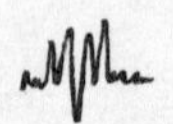

Less than a decade ago, prognosticators about the "wonders" of the digital revolution predicted the widespread use of 3-D "virtual reality" headsets—a brave new world where even the thirty-second sound bite would be obsolete, "reality" itself replaced by a stereopticon of flashing CGI images and surround sound, a simulacrum of life, our own reality bubble geared mostly to sensation-driven entertainment (purveyors of pornography were over the moon). Today, these same prognosticators have to contend with the awkward fact that the most momentous outcome of the "Digital Renaissance" has been (as reformed digital evangelist and now leader of a movement called "Team Human," Douglas Rushkoff, has gleefully pointed out) . . . *the rebirth of radio*. Virtual reality headsets have gone the way of Google glasses. By far the most popular use to which people put their iPhones, Rushkoff points out, is the downloading of three-hour podcasts by the likes of Joe Rogan or Sam Harris—podcasts that often feature knotty and complex discussion of politics, science, philosophy (punctuated, in Rogan's case, by extended musings on Mixed Martial Arts). When not listening to podcasts, we are bingeing TED talks on equally complicated subjects on YouTube. Both forms are distinguished by their singular focus on one of the oldest technologies on Earth: the human voice, isolated, talking.

In 2000, I was invited to participate in a quaint-sounding event in downtown Manhattan run by an organization calling itself The Moth—they wanted me to come and "tell a story," and they assured me that there would be a decent crowd. And to my surprise there was; easily a hundred people, maybe more, crammed the small bar. The crowds have gotten bigger since. The Moth storytelling events have spread around the world, in venues that hold thousands and according to The Moth's website their podcast (featuring nothing but single voices telling stories) is downloaded over fifty million times a year, their *Moth Radio Hour* is broadcast on over five hundred radio stations across the planet. This human voice thing is *big*!

Our passion for the sound of the human voice in marathons of performative speech has also led to a boom in the sale of audiobooks, which had never really caught on when first introduced in the 1970s as mail-order boxes of ugly cassette tapes. Digital download and streaming changed all that over a decade ago. Nevertheless, I was resistant (convinced that no narrator could surpass the one I carried around in my head)—until 2009, when I had to master an interview subject's massive literary oeuvre in a very short time and realized that, to do it, I would have to use every available moment of the day—including those when traditional reading was impossible (jogging; grocery shopping; doing dishes). To my surprise, I learned that the audiobook was a better, and more efficient, and even more pleasurable way to absorb written material than silent reading. The best audiobook narrators made fiction and nonfiction come alive in ways that my own inner voice and ear often could not match. But the truly revelatory realization was that (counterintuitively), *listening* to a book aided my comprehension and retention. Amazed, I emailed a leading neuroscientist, V. S. Ramachandran, and asked whether there existed some brain-based reason audiobooks are so effective as prose-delivery devices. In response, he pointed out that our language comprehension and production evolved in connection with our *hearing*, around 150,000 years ago. Writing is only 5,000 to 7,000 years old—"partially going piggyback on the same circuits," he wrote. "So it's possible LISTENING to speech (including such things as cadence, rhythm, and intonation) is more spontaneously comprehensible and linked to emotional brain centers—hence more evocative and natural."[33]

I like this explanation, not least because it accords with what I have since learned about how so much of what we find effective, moving, pleasurable, illuminating, and persuasive in *written* texts had its origins in the *spoken* rhetoric of Cicero and the other ancient writers on oratory. I also like it because it means that our hunger for, and love of, the human voice can never become obsolete, or *outmoded*: it is simply too much a *part of us*, a part of our neural circuitry, a primary means by which we make sense of the world, of how we interpret reality. This is why, in democratic societies, the sound of the human voice is so intimately in-wound with how we pick the people we choose to lead us. I would even argue, with Cicero and the other ancient philosophers of rhetoric and oratory, that the voice is the *primary* means we use to decide for whom to vote.

When responsibly, honestly—and honorably—deployed, the sound of a single human voice speaking is still the best index to the worthiness of a leader. The best do not talk to us in dumbed-down sound bites or focus-grouped slogans; are not content with tailoring every address to the formats and time constraints of twenty-two-minute "Nightly News" broadcasts, or four-second GIFs posted to Twitter. They do not rail as demagogues. Instead, they address a nation, and world, in language that assumes intelligence, stamina, patience, and reason in their audience. Admittedly, such leaders are rare at any time in human history. People alive today over the age of, say, ten, have the privilege of having heard one of them.

Barack Obama first burst on the global consciousness in 2004 at the Democratic National Convention, where he delivered a keynote speech that expressly addressed the nation's growing political, racial, and class divisions—and, in the best Ciceronian fashion, sought to bridge them. An Illinois state senator who had begun life as the son of a Black, Kenyan-born father and white mother from Kansas, Obama, then forty-two years old, cast his own improbable rise as proof of how exactly such divisions can be bridged, explaining that his own presence on the stage was proof of the promise that American democracy offers every citizen. "In no other country on earth," he said, "is my story even *possible.*" But, he warned, that promise could be fully realized only if the country remembered its motto: *e pluribus unum*—"out of many one." No addle-pated optimist, Obama warned his audience that, even as he spoke, "there are those preparing to divide us." But, he continued, "I say to them tonight"—here, his voice began to swell to a dramatic crescendo, his rhetoric taking on the forms endorsed by Cicero centuries ago, including a rhythmic repetition of a certain word accentuated by a rising volume on each reiteration—"there's not a liberal America and a conservative America—there's the United States of America; there's not a BlackAmeri*ca* and a white Ameri*ca* and Latino Ameri*ca* and Asian Ameri*ca*! There's the United *States* of Ameri*ca*!"

There was emotion in Obama's voice—the ancients had never denied a role for emotion in persuasive oratory, indeed they had said it was crucial to *move* audiences—but it was not the emotion of the demagogue: it was emotion acting in the service of emphasizing complex, underlying ideas—ideas of equality and unity enshrined in the Constitution, which Obama had taught as a professor at the University of Chicago Law School. The stirring "melody" of his voice thus operated in service of the "lyrics," which reminded his listeners of the fragile ideal upon which the founders had launched the experiment of America.

The music of Obama's voice also embodied the very ideal of oneness-from-diversity that he was describing in his keynote speech, since that voice, first shaped in its basic prosodic stress patterns from the acoustic signal that reached him, in the womb, from his white Kansan mother, was overlaid with linguistic influences absorbed during an astonishingly diverse childhood and youth, which included years spent in Seattle, Jakarta, Honolulu, Los Angeles, and New York. The result was a voice that was, at once, as General American in accent as the most familiar TV news anchor, but with the subtlest African American touches (or "frills") on some vowels—particularly the long *ee* sound at the ends of multisyllabic words like "history" or "presidency"[34]—as well as clear echoes of the fiery

Black gospel church oratory, the "tonal semantics," that he had heard, and internalized, as a regular attendee of the Trinity United Church of Christ in Chicago, a Baptist congregation of which he had been a member for twenty years. The echo of Black church oratory was especially audible in the peroration of his 2004 DNC speech, when he hit an MLK-like crescendo after a (very MLK-styled, and Cicero-derived) repetition of the word *America*—which he pronounced with an unusual emphasis on the *final* syllable—"Ameri*CAH*"—an artful oratorical choice which forced his listeners to, as it were, hear the word anew.

Four years later, during his run for the presidency, Obama revealed a virtuosic control of his vocal instrument that that masterful 2004 keynote speech barely hinted at—both on, and off, stage. In diners and churches, veterans halls and country fairs, debate stages and large rallies—speaking to crowds of every racial background, age, sex, educational level, and class—he demonstrated an ability to code-switch seamlessly between his various vocal registers: that of the crisply enunciating, all-business Chicago Law School professor, but also that of the easygoin', *g*-droppin' Everyman (a study by Labov revealed that, counterintuitively enough, Obama dropped his *g*'s as often as rootin' tootin' Sarah Palin),[35] as well as the Columbia University undergrad who had lived on West 109th Street and, upon graduation, moved to East 94th Street, which he described in *Dreams from My Father* as "that unnamed, shifting border between East Harlem and the rest of Manhattan." There, he absorbed the intonations, accent features, and grammar of the inner city's Black population. "Nah—we straight," he famously called out at Ben's Chili Bowl in Washington, D.C., when the Black counterman tried to hand him his change—a pitch-perfect use of Black English, from the fronted *o* vowel that turned "No" into "Nah," to the omitted copula in "we straight," to the street lingo for "even" ("straight").[36]

Obama's skill at code-switching was, political analysts agreed, essential to his negotiating an electorate at once suspicious that he might, as Obama himself put it, be *too Black*, or *not Black enough*. While critics suggested that this vocal shape-shifting marked Obama as a fake, David Remnick, in his 2010 biography of Obama, *The Bridge*, pointed out that this was not a "cynical gift," but instead the legitimate legacy of Obama's rich and varied life background, as well as a talent he shared with Martin Luther King Jr.[37] Obama himself said that he really didn't give much thought to how he slipped "into a slightly different dialect" with Black and white audiences. "There's a level of self-consciousness about these issues that a previous generation had to negotiate that I don't feel I have to," he said.[38] In short, the sounds that sprang to his lips in spontaneous speech, conversation, or in the heat-of-the-moment at the podium originated, like everyone else's, in myelinated brain circuitry over which he had little control—or felt no special need to *exert* control.

In his 2008 presidential primary campaign, Obama made history when, in defeating fellow Democrat Hillary Clinton, he became the first African American to win a major party's nomination—but not before he was obliged to jump a final, perilous, oratorical hurdle. Impolitic utterances by Rev. Jeremiah Wright, the pastor of Trinity United, led to charges, by political opponents, of a Black nationalist extremism hiding behind Obama's affable exterior. At first ignoring the preposterous charge, Obama finally had to address it when repeated airings of Wright's comments on twenty-four-hour cable news channels threatened to capsize his candidacy. The speech that he delivered, "A More Perfect Union," on March 18, 2008, at the Constitution Center in Philadelphia, is second only to King's "I Have a Dream" speech, of which it was a kind of bookend, and a masterpiece of American oratory on a par, I would contend, with the greatest of Lincoln's addresses. By reminding the nation of its founding principles, it showed that the only path forward, as a democracy, was to adhere to those principles. And this is to say nothing of the impossibly high demands on Obama's vocal performance, on his oratorical delivery, that might be unmatched in the history of American electoral politics. A single overemphasis on a syllable by a too-taught tension on the vocal cords—a single misjudgment on how much lung power to use in driving up the volume in a phrase—could have, *would* have, convicted him of the charge of being the *dangerously angry Black man* that had occasioned the speech in the first place. That Obama turned the moment into so much more than a defense of his own candidacy is where the speech's true greatness lies.

Cicero said that great oratory draws on deep knowledge of language, literature, history, politics, and psychology. He also said that it cannot help but reveal the speaker's true self, the content of his character (if you will), even as it unites an audience with a sense of shared purpose. Obama's speech in Philadelphia did all of those things as he explained to America, in tones both resigned and passionate, the original sin it had committed against not only the Africans it had kidnapped and placed into slavery, but against the foundational ideal that so inspired Abraham Lincoln: that *all people are created equal*—and how that sin echoed down the years to the present moment. Obama did not let racists off the hook, but he also did not demonize all white Americans. He denounced the "incendiary language" of his former pastor, whose views threatened "to widen the racial divide," and "rightly offend white and Black alike," but he refused to renounce altogether the pastor who had strengthened Obama's faith and worked to alleviate the sufferings of the poor and sick. Obama even spoke to the legitimacy of the anger within those "working- and middle-class white Americans" who do not feel "particularly privileged by their race," who "see their jobs shipped overseas or their pension dumped after a lifetime of labor," and who have "come to see opportunity as a zero-sum game, in which *your* dreams come at *my* expense."

From a moment that seemed designed to force Obama into a divisive stance, he somehow turned the occasion into his most eloquent and passionate appeal for unity, understanding, and tolerance. Whatever else can be said of him and his presidency, Obama is indisputably an exemplary *person*: decent, upright, intelligent, and honest in his desire to solve class divisions and foster the racial unity inscribed in his own DNA—virtues made audible that day in Philadelphia, in both his words and the voice that carried them, and which helped to earn him, finally and historically, the office of president of the United States.

In neuroscientific terms, Obama put the emotional channel of his voice—the paralinguistic, prosodic signals that emerge from the brain's limbic centers—in the service of his higher, executive brain: the cortex, where ideas, reason, and language are generated. Everyone communicates, vocally, from these layers of the brain, blending "music" and "lyrics" in a manner that most persuasively conveys an idea or emotion. It is the demagogue (in public speech) or the abusive bully (in private) who privileges the limbic brain above the cortex, who tunes his voice to atavistic roars and growls, gasps and shrieks, who weaponizes the voice in ways that appeal primarily to the listener's emotion centers thereby activating her primal, irrational, animal instincts of fear, envy, anger, resentment, vengeance. This is how the demagogue, although unfit for office, attains electoral victory and, as leader, assumes the role of tyrant or dictator, eradicating the institutions of democracy that would rein in his power, while continuing to prey on the fears and anger of the populace with his vocal attack—to move them, in the worst-case scenario, to acts of barbarity that strain credulity. This is what occurred in Germany with the rise of Adolf Hitler, the most destructive demagogue in human history.

Hitler himself recognized the essential role that his voice played in propelling him to power. A failed artist and chronically unemployed tramp, he was a thirty-year-old First World War army veteran—a disaffected loner inwardly seething with anti-Semitic and anticommunist fervor—when he attended a meeting on September 12, 1919, of the German Workers Party, one of countless antigovernment groups in desolated postwar Germany. Unimpressed by the collection of twenty or so shabby discontents he found in the murky basement of Munich's Sterneckerbräu beer cellar, Hitler was making a hasty exit when he became incensed by the comments of a "professor" who dared to air an opinion that contradicted Hitler's own. Despite his usual diffidence among strangers, Hitler found himself roaring a torrent of abuse so violent and sustained that the professor fled, as Hitler later wrote in *Mein Kampf*, "like a wet poodle."[39] Meanwhile, the group's leaders looked on with "astonished faces," riveted by the annihilating force of Hitler's voice. They immediately invited him to start giving public speeches on behalf of the group. He agreed and joined the armies of disenfranchised cranks who railed from soapboxes, flophouses, beer cellars, soup kitchens, and street corners across Germany. The difference with Hitler was that he had a voice people could not ignore—a voice infused with a ferocious passion that was both a calculated oratorical choice, and something beyond Hitler's control: his real and true, poisonously paranoid, racist self escaping into the open.

That Hitler was acutely conscious of the power of a single voice to foment mass political movements is clear from *Mein Kampf*:

> The power which has always started the greatest religious and political avalanches in history rolling has from time immemorial been the

> magic power of the spoken word, and that alone. The broad masses of the people can be moved only by the power of speech. All great movements are popular movements, volcanic eruptions of human passions and emotional sentiments, stirred either by the cruel Goddess of Distress or the firebrand of the word hurled among the masses.[40]

Hitler's first chance to hurl his own voice among the masses came one month after that meeting in the Munich beer cellar, when he made his debut speech on behalf of the German Workers Party (soon to be renamed the National Socialist, or Nazi, party). He spoke for thirty minutes to 107 people, but he tore into his text with a spittle-spewing, arm-flailing, larynx-ripping, fist-shaking passion, stunning his audience, and providing, for Hitler, an epiphany: "what before I had simply felt within me, without in any way knowing it, was now proved by reality: I could speak!"[41]

What he spoke *about* were ideas that had been festering within him ever since Germany signed the Treaty of Versailles, which had ended the First World War and mandated that Germany hand back the territories it had seized, dismantle its army, and pay massive war reparations. Many Germans, guilt-ridden, demoralized, and humiliated by the country's abject defeat, were prepared to accept the terms. Hitler was not. Animated by a militaristic mania born in the trenches, and the conviction that Germany not only had nothing to apologize for but still deserved world domination, he blamed the country's "humiliation" on the weakness of its leaders, and found a scapegoat for Germany's financial woes in a group he had long demonized: Jewish bankers and, by extension, the Jews at large.

Over the next two years, Hitler shrieked, wailed, growled, wept, whimpered, and bellowed his way through speeches that, rhetorically, were almost devoid of logical argument—save for their simple, slogan-like appeals to hatred, anger, recrimination, self-pity, and a fervid German nationalism and Aryan "racial purity." When a collection of Hitler's speeches, *My New Order*, was translated into English by Raoul de Roussy de Sales, in 1941, it carried a preface in which de Sales emphasized the crudity and awkwardness of Hitler's language, and the peculiar violence with which he put it across. "Hitler's speeches are weapons," he wrote, "as much a part of his strategy of conquest as more direct instruments of warfare." Filled with lies and contradictions, they nevertheless produced "the effect of a psychological pincer-movement which crushes the best defenses of logic and ordinary morality."[42] They did this, de Sales said, through the raging power of Hitler's voice, and his brutal repetitiveness. Like a mortar dispelling round after round of ear- and will-shattering shells, Hitler exploded the same simple, deranged, ideas, over and over and over again, pounding "into the heads of his listeners the same formulas."

Joseph Goebbels, the head of Nazi propaganda, agreed that *what* Hitler said was immaterial; it was *how* he said it. In 1936, Goebbels wrote about Hitler's voice and its power to "reach out from the depths of his blood into the depths of the souls of his listeners," to "rouse the tired and lazy, fire up the indifferent and the doubting, turn cowards into men and weaklings into heroes."[43] At no point did Goebbels say that Hitler achieved these effects through the precision of his language, or the rationality of his ideas. Quite the contrary. "The important thing is not whether an idea is right," Goebbels wrote, "the decisive thing is whether one can present it effectively to the masses so that they become its adherents"—and that was done through the power of Hitler's *voice*. "The magic of his voice reaches men's secret feelings," Goebbels wrote. "There is probably no educated person in the world who has not heard the sound of his voice and who, *whether he understood the words or not* [my italics] felt that his heart was spoken to."[44]

A special division of British intelligence dedicated to monitoring Hitler's speeches for signs of his mental state formed its own theory of Hitler's voice during the Second World War. Noting that he tended to begin each address in a quiet register that built to a sustained crescendo, British intelligence believed that Hitler, in his fitlike climaxes, entered a "trance" that resembled that of a Shaman—the healers and medicine men of indigenous tribes who transmit "messages from the spirits" through sustained, hypnotic chanting. One intelligence report suggested that the epilepsy-like neural firing in Hitler's brain was transferred, through his vocal sound waves, to his listeners' brains where it touched off a symmetrical neural fit, synchronizing them into an entrained state of rage and blood lust.[45] Others explained the contagious fury of Hitler's voice in purely psychological terms, focusing on the depressed state of the German people before Hitler's rise—a state of defeat and despair that predisposed them to react to the rousing stimulus of a demagogic voice whose sonic furor carried a single message: *Germany will rise again*.

One thing is certain: the sound of that voice sparked an extraordinary reaction in the populace that was reflected in the vertiginous growth of Hitler's crowds: a year after his maiden speech to just over 100 people, he spoke at the Zirkus Krone, the largest hall in Munich, to an audience of 5,600; nine months later, he held a Nazi rally that drew 14,000; two weeks after that, he drew 20,000. These rallies would often degenerate into violent mosh pits as his followers, coaxed to frenzy by his voice, fought each other and Hitler's stormtrooper security detail.

To these live audiences were added the millions Hitler reached through the radio. The French-American novelist George Steiner was a child when Hitler consolidated his hold over Germany in the 1930s. Steiner told author Ron Rosenbaum about the "spellbinding" power of Hitler's voice as it leapt over the airwaves and into the room: "my earliest memories are of sitting in the kitchen hearing the voice on the radio.

It's a hard thing to describe, but the voice itself was mesmeric. . . . The amazing thing is that the body comes through on the radio. I can't put it any other way."[46]

Hard as it is to comprehend in retrospect, Hitler first came to power in Germany through normal democratic electoral processes: in 1932, he ran against General Paul von Hindenburg in the presidential elections, and although he came in second, he garnered nearly 37 percent of the vote—a remarkable tally for a man who preached murderous hatred against anyone but Germans of pure "Aryan" blood. Up to this point, Germany's intelligentsia and business elite treated Hitler as a buffoon and joke. No more. Seeing the fevered following of millions that Hitler had built up, business leaders pressed Hindenburg to make room for the Nazi party within the government. Hindenburg reluctantly agreed and appointed Hitler chancellor. Four weeks later, on February 27, 1933, the German parliamentary building, the Reichstag, was mysteriously lit on fire by unidentified terrorists (many historians believe the fire was set on Hitler's orders). This triggered a national panic like that which followed the 9/11 attacks in America. Hitler railed that the arson was by *communists* and *Jews* bent on overthrowing the government, and convinced Hindenburg to suspend basic civil rights, allowing mass detentions without trial. Some four thousand communists, and suspected communists, were locked up.

In new elections, a month later, the Nazi party secured still more of the vote: 43.9 percent—enough for Hitler to seize control. Using the very structures of democracy to dismantle the country's democratic institutions, Hitler claimed that the communist-terrorist-Jewish risk to Germany required desperate measures, and he wrote new laws without the consent of parliament, a dramatic and unprecedented departure from the constitution. Now, on Hitler's say-so, rival parties could be *legally* shut down. They were—and their assets seized. Hitler's stormtroopers, acting legally, destroyed union offices across Germany, and their leaders were interned in the concentration camps hastily being constructed across the country. In July, Hitler declared his Nazi party the only legal political entity in Germany, which empowered him, lawfully, to go after his political adversaries. They were rounded up, arrested, and shot.

Hitler seized control of all opposition newspapers and radio stations and arrested their owners, silencing all dissent. When Hindenburg died in July at age eighty-five, Hitler became both head of the country and the armed forces. He soon launched his war of global conquest—a war that would result in defeat for Germany and the deaths of some 75 million people, soldier and civilian, including a genocide against six million Jews: a staggering Holocaust that had begun, improbably enough, in a smoky Munich beer cellar with an outburst that revealed to Hitler an unsuspected political gift: his voice. As Raoul de Roussy de Sales wrote in the preface to *My New Order*, "he started as a soap-box orator and spoke his way to power."[47]

The horrors Hitler unleashed seemed to alert the world to the dangers of demagoguery, populist movements, and dictatorship. Over the ensuing half century, democracy spread around the world, helped indeed by oratory like JFK's *Ich bin ein Berliner* speech. But at the dawn of the new century, political scientists noted a worrying global trend: democracies were beginning to weaken and fail. Since 2010, populist political parties with charismatic, demagogic, antidemocratic leaders espousing anti-immigrant rhetoric, promoting nationalism and de-globalization have gained power and influence throughout Europe—in countries including Poland, Hungary, Austria, the Netherlands, and France.[48] Likewise in Brazil and India. The root causes of this lurch from democracy are consistent across all nations: an erosion of living standards for the middle class through growing concentration of wealth in the hands of a tiny minority of the rich (the very trend FDR had turned back with the New Deal); the increasing multiethnic makeup of countries (which allows for the whipping up of racist sentiment against the other); and a declining legacy media with the rise of the internet and the easy spread of propaganda and fake news.

When political scientists poll people on their attitudes to democracy today, only one-third of Americans in their twenties and thirties say it is "very important." In Sweden, a chilling 26 percent of eighteen- to twenty-nine-year-olds say that having one strong leader would be better than bothering with elections every few years.[49] In June 2016, England was swept up in the global populist wave, through an emotional call to nationalism, anti-immigration, and economic self-determinism by the cynical, dishonest, demagogic oratory of Tory Boris Johnson, UK Independence Party (UKIP) leader Nigel Farage, and House of Commons leader Jacob Rees-Mogg. To the world's shock and surprise, England voted to leave the European Union, with Brexit.

Even with the dangers of demagoguery striking so close to home, America's political and pundit class continued to declare that *it can't happen here*, and in the months, weeks, and days leading up to the November 8, 2016, presidential election, pollsters and other "experts" confidently predicted that the winner would be Democrat Hillary Clinton. Hours into election night itself, the *New York Times*'s now-notorious "election needle" indicated that there was a better than 80 percent likelihood of Clinton becoming the first woman president of the United States.

A little more than one year and four months earlier, on June 16, 2015, Donald Trump, a sixty-nine-year-old failed real estate developer, bankrupted casino magnate, and star of a declining reality television show, mounted a stage in the lobby of the midtown Manhattan skyscraper bearing his name, adjusted the microphone on the podium, and announced his candidacy for president. What everyone would remember was his offensive claim that Mexicans who come to America are "bringing drugs, they're bringing crime, they're rapists. And some, I assume, are good people"—the most nakedly racist statement uttered by a political candidate in recent American history. But the volume and pitch at which Trump blared this piece of hate speech was almost as jarring as the statement itself. His voice cut through the air like a cabbie's horn on Fifth Avenue. At least to some ears, it was a voice instantly disqualifying as leader of the free world, and not only in terms of the malignant ideas it bodied forth.

For one thing, it was too high. Multiple studies of voice and electability have shown that speakers with a higher pitch consistently lose to candidates with a lower fundamental frequency[50]—a finding applicable also to women when going head-to-head against each other in an election. And Trump, despite being relatively tall at six-foot-two, has a notably high voice, indeed several Hertz higher than average,[51] which sends a signal not of dominance but submission. This evolutionary "disadvantage" actually explains a peculiarity that has been seized on by every comedian attempting to impersonate him: namely, the unusual way that he purses his lips and pushes them outward as he speaks. The only credible explanation that I have come across for this highly unusual articulatory habit are findings from a scientific paper coauthored, almost five decades ago, by Chomsky's linguistic nemesis, Philip Lieberman.

In the early 1970s, Lieberman studied why prepubertal boys and girls can, in many instances, be told apart by the sound of their vocal pitch even though they possess vocal cords and resonance chambers of equal size, and thus should sound identical. He discovered that many prepubertal boys, in an unconscious bid to lower their voice and distinguish themselves from girls, round and extend their lips while talking, increasing the overall length of the vocal tract, boosting the lower frequencies in the voice spectrum, and giving the illusion of a deeper voice.[52] For a person like Trump, so consumed by the need to dominate and be the alpha male in every circumstance, it seems likely that—sometime after puberty and upon his perceiving that he possesses a less dominant voice than other males—he intuitively hit on the expedient of rounding and pushing out his lips to lower his pitch slightly.

Excessive volume was another way that Trump learned to overcome the innate disadvantage of his "submissive" pitch. The blaring, shouting register that he used when announcing his candidacy would characterize all his subsequent public appearances as candidate and later president—even when not standing beside an idling helicopter.

Blaring volume is an instinctive reflex for weaponizing the voice, and one exaggerated by blowhards and domineering personalities everywhere, but Trump's use of this approach also derived from a most unlikely source: the years he spent in the 1980s and 1990s playing host, at his Atlantic City casinos, to the WWE professional wrestling championships.[53] There, Trump absorbed important lessons about how to whip up popular support from exactly the disaffected, blue-collar, Rust Belt demographic that would enthusiastically (and decisively) pull the lever for him in the election. Although professional wrestlers are supposedly athletes and fighters, the ones that become most popular with the public—who "get over," in wrestling parlance—are those who dominate in the *verbal* smackdowns at ringside. Hulk Hogan, by no means the most skilled in the ring, nevertheless became the most famous, and well-paid, wrestler of his era thanks to his *shouting matches* against opponents. Trump's vocal style is pure pro wrestling.

At first, the media treated Trump's candidacy as a joke: television covered it as a ratings-grabbing car crash, and online platforms as clickbait.

The Huffington Post put all Trump coverage in their Entertainment section. ("Our reason is simple. Trump's campaign is a sideshow.")[54] But six months later, Trump was the front-runner for the Republican nomination and feeling emboldened. On December 7, 2015, he held a press conference in which he called for a "total and complete shutdown of Muslims entering the United States"—a proposal that smacked too much of an earlier demagogue's demonization of an entire religion. Arianna Huffington announced that her site was shifting Trump coverage from the Entertainment to the Politics section owing to Trump's "vicious pronouncement," which showed that his campaign had "morphed" into "an ugly and dangerous force."[55]

In reality, the ugliness and danger had been there for all to see for decades. In 1973, Trump and his father were charged, by the Department of Justice, with discriminating against potential tenants on racial grounds.[56] In 1989, when five Black and Hispanic teenagers were arrested and railroaded into confessing to the beating of a woman jogging in Central Park, Trump took out a full-page ad in the *New York Times* calling for their summary execution.[57] (DNA evidence and a confession by the actual perpetrator later exonerated all five suspects, but Trump, during his presidential campaign, said that he still believed them guilty.)[58] In 1990, it emerged, from the divorce deposition of Trump's first wife, Ivana, that Trump kept, beside the marital bed, a copy of Hitler's speeches: *My New Order*—the collection translated by Raoul de Roussy de Sales. "*If* I had these speeches," Trump told Marie Brenner of *Vanity Fair*, "and I'm not saying that I do, I would never read them."[59]

In *Fear*, a chronicle of Trump's political ascent and first year in office, author Bob Woodward says that Steve Bannon, in agreeing to work on Trump's campaign, did so only after careful evaluation of Trump's advantages over Hillary Clinton. It all came down to their voices. "He spoke in a voice that did not sound political," Bannon told Woodward. "Her tempo was overly practiced. Even when telling the truth, she sounded like she was lying to you." Hillary's voice, Bannon added, came "not from the heart or from deep conviction, but from some highly paid consultant's talking points—*not angry.*"[60]

As shockingly angry as Trump's voice sounded, it was not unprecedented in modern presidential politics. In 1992, Pat Buchanan, a former special consultant to Richard Nixon, had mounted a primary challenge for the Republican nomination against George H. W. Bush. Although never descending to Trump's simian bellowing, Buchanan spoke in snarling, pugilistic tones that, at the time, were startling. But Buchanan's effort to whip up a populist movement founded on splitting America along divisions of race, income, and "cultural values" got nowhere. America was, at the time, entering one of the longest economic booms in history, the country was not at war, and the terrorist attacks of 9/11 were years away. Cicero would have known that it was not an opportune moment for a demagogue.

But by the time Donald Trump faced off against Hillary Clinton, America was in very different condition. A Republican-controlled House and Senate had blocked every Obama initiative, resulting in political paralysis in Washington. Large segments of the Midwest had been decimated by the decades-long flight of factory jobs to the developing world: unemployed factory workers had succumbed to the anger and resentment that Obama had warned against in his "More Perfect Union" speech, when he beseeched Americans not to see "opportunity as a zero-sum game, in which *your* dreams come at *my* expense." Trump did the precise opposite. In campaign rallies that filled 2,000-, 3,000-, 8,000-, and 20,000-seat arenas, Trump urged out-of-work white Americans to see their predicament as the fault of Mexican immigrants and other dark-skinned interlopers. As a solution to terrorism, he repeated his call for banning Muslims from entering the country. Rather than try to heal the growing partisan divide between Democrats and Republicans, he fired up red state voters by vowing not only to beat Hillary Clinton, but to jail her. Sacrificing reasoned argument to brief, quick-fix slogans that could be roared from the podium, and echoed back to him in unison by the thousands who packed his rallies, Trump led his followers in chants of "Build the Wall!" "Lock Her Up!" "Make America Great Again!" His stadium rallies became cauldrons of violence. When a heckler at a rally in Las Vegas was manhandled by Trump's supporters, Trump shouted from the stage: "You know what they used to do to guys like that when they were in a place like this? They'd be carried out on stretchers, folks!" To the roars of the crowd, he added: "I'd like to punch him in the face."[61] At a rally in Cedar Rapids, Iowa, Trump told the crowd how to handle protesters: "Knock the crap out of them, would you?" he said, to a cyclone of cheers and laughter. "I will pay the legal fees, I promise you."[62] At a rally in Wilmington, North Carolina, Trump hinted at a more draconian method for dealing with his Democratic rival than merely locking her up. "If she gets to pick her judges, nothing you can do folks," Trump said. "Although the Second Amendment people—maybe there is, I don't know!"[63]

In healthy democratic societies, a free press serves as the best check on demagoguery, but no more. Trump had first joined Twitter in March 2009—and he quickly understood that it had replaced old media as the prime tool for mass influence and persuasion ("It's like owning your own newspaper—without the losses," he tweeted in November 2012). As a candidate, Trump used the platform as an eerily exact extension of his corporeal voice: his tweets—often typo-riddled, in all caps, and with multiple exclamation points—were issued like the vocal blows he used to pummel opponents into silence. "Hillary will never reform Wall Street. She is owned by Wall Street!"[64] "#WheresHillary? Sleeping!!!!!"[65] "AMERICA FIRST!"[66] Meanwhile, he exhorted his followers to ignore what the mainstream press was saying about him. They were "fake news" and (in a phrase borrowed from Goebbels, who'd used it to demonize the Jews) "the enemy of the people."

The frustration on the part of old media entities to act as a check on Trump was palpable. In November 2017, ten months into Trump's presidency, MSNBC host Lawrence O'Donnell fumed over how Trump and his surrogates could continue to lie and contradict themselves without facing any consequences. In this particular instance, O'Donnell was incensed by comments from Bannon, who had performed a whiplash-inducing 180-degree turn on a statement from just hours before. People must be "fools," O'Donnell sputtered, not to see that Bannon, like Trump, is a "hustler" and "liar."

"But he's not *exactly* a liar," said O'Donnell's guest, pundit David Frum, who understood perfectly how Trump and those around him used their voices. "Even for a liar, words have *meaning*," Frum said. "Think of it as *music*. He's a *musician*. These are *tones* of grievance and anger. And it is an important thing to understand because we're going to hear more of this in the coming months and years."[67]

The best explanation of how that music of "grievance and anger" won the support of over sixty million Americans is found in a book written before Trump even announced his candidacy (although published afterward, in 2016): *Hillbilly Elegy*, a memoir by J. D. Vance of growing up in the Rust Belt city of Middletown, Ohio, in the 1990s and early 2000s. A once thriving steel town, Middletown had, like so many manufacturing cities, been destroyed by the offshoring of its factories and the subsequent widespread unemployment that led, first, to creeping despair, then to alcohol and opioid addiction, and finally to the loss of all hope.

Middletown is a working-class white population suspicious, Vance wrote, of "outsiders or people who are different from us, whether the differences lie in how they look, how they act, or, *most important*, *how they talk*" (my italics).[68] This hypersensitivity to the vocal channel, and its in-group/out-group signaling (which would have surprised Labov not at all), extended to the person who occupied the White House during the time that Vance was writing: Barack Obama, about whom Middletownians felt unease and even dislike—emotions born not of racism, Vance insists, but of the understandable envy they felt for Obama's success, education, and healthy, happy family. Nothing made Obama seem more "alien" to the despairing citizens of Middletown, Vance wrote, than his voice. A thing of inspiration to those Americans fortunate enough to share Obama's prosperity and prospects, that voice was, to people in Middletown, his most off-putting characteristic, precisely because it embodied everything admirable in the man, everything *they* had lost or never had. "His accent—clean, perfect, neutral—is foreign," Vance wrote.[69]

It is impossible to read these passages about Obama, and about the depths to which Middletownians had sunk, and still wonder why they, and millions of other Americans across the Rust Belt and other failing parts of America, were galvanized by the sound of Donald Trump's venomous, vengeful voice.

Since Trump's victory, America has learned some harsh lessons about what happens when a leader seething with prejudice and anger uses the mechanisms of democracy to seize the national microphone. His voice amplifies voices like his own. On the very day of Trump's electoral triumph, former grand wizard of the KKK David Duke, who had not been heard from in the mainstream media in over thirty years, was suddenly once again all over television screens, voicing his delight at Trump's election—and tweeting to Trump: "We did it!" Mediagenic white supremacist Richard Spencer was extensively interviewed on all the major television networks and profiled in magazines and newspapers. On August 11, 2017, seven months into Trump's presidency, the world was treated to the scarcely believable sight of hundreds of torch-bearing

American neo-Nazis who had converged on the city of Charlottesville, Virginia, to stage a "Unite the Right" rally. Marching through the city, faces brazenly exposed to cameras as if they no longer had anything to fear in the way of retribution from employers or anyone else, they chanted "Jews will not replace us" and "Blood and Soil"—phrases drawn from de Sales's English translation of Hitler's speeches.

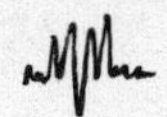

Given how faithful an index the human voice is to the thoughts, disposition, personality, intelligence, and judgment of a human being, there is every reason to feel fearful when someone like Donald Trump becomes the most powerful person in the world. The very sound of his voice—apart even from the ugliness of his utterances—announces his low impulse control and roiling anger, his tendency to reach for rash and oversimplified solutions to complex problems, his animal need to dominate—to "be a killer" (as his father put it to him in childhood).[70] One hesitates to imagine how such a man might react if called upon to exercise the nuanced decision making and restraint demonstrated by John F. Kennedy during the Cuban Missile Crisis, qualities of mind and temperament uniquely demanded of the leaders of superpowers in an age of thermonuclear weaponry.

Nor is nuclear Armageddon the only nightmare scenario that this perilous moment calls to mind. As I write, the leaders of every country in the world are grappling with how to stem a pandemic, the coronavirus—a crisis that demands global cooperation, coolheadedness, strategic reasoning, and the ability for decision makers to look beyond their own personal advancement. The president of the United States initially dismissed fears about the virus as a "hoax" perpetrated by a liberal mainstream media, working in collusion with a Democratic House of Representatives, to sabotage his reelection chances by cratering the stock market and depressing the economy. Precious weeks passed in inaction as the virus took hold across the United States, another sobering reminder of the dangers that the ancient Greeks warned of in democracy—a system where a person, no matter how unfit, can speak his way to power and, through misuse of that power, threaten to bring our species, and every other, to extinction.

EIGHT

SWAN SONG

A vision of our annihilation in a nuclear conflagration or unchecked pandemic is too grim a note upon which to end this tribute to the gifts that we give, and are given, through the voice. Arguably, the greatest of those gifts is one we've only glanced at thus far: song, which I have addressed primarily in connection with Darwin's insight that language began with singing. For some, however, the musical origins of speech are precisely why singing is a nonessential subject of inquiry into what made us human. Steven Pinker famously states that the melody and rhythm that early humans recruited for language render music a topic of secondary interest, at best, since these prosodic features provided nothing like the decisive evolutionary advantages that words and grammar did. We respond so powerfully to music, Pinker says, simply because it exaggerates the musical channel of speech, widening the swoops in melody, accentuating the beats and rhythms of talking, and thereby pushing harder on the buttons in the brain's pleasure centers that early humans recruited for deciphering language. Thus, Pinker's description of music as "auditory cheesecake"[1]: delicious for its artery-clogging overabundance of fats and sugar (or melody and rhythm), but hardly essential to life. Pinker concludes that "music could vanish from our species and the rest of our lifestyle would be virtually unchanged."[2]

For anyone (like myself) who draws such pleasure and solace from both making and listening to music, it is hard to credit this view, but to be fair to Pinker, it is also hard to explain the peculiar power and importance to our existence of music and song. President Obama inadvertently hinted at it, in June 2015, when he led a memorial service for the nine African Americans slaughtered at a church in Charleston, South Carolina, by a Glock-wielding white supremacist. A minute or two into his spoken remarks, Obama paused for an extraordinary twelve seconds. When he broke the silence, it was with the opening note of "Amazing Grace." That one of our most eloquent orators found the medium of speech inadequate to that moment of shared anguish was telling. Whatever else Obama's rendition of "Amazing Grace" was, it was not auditory cheesecake. It was far more nourishing (emotionally, psychologically, and, in the parlance of faith, spiritually), and it proved the mysterious power, and fundamental *necessity*, of singing when words alone fail us.

The first formal inquiries into the power of song were made in the 1920s by researchers using a then-revolutionary technology for capturing the voice and rendering it *visually*: phonophotography, which translated the vocal sound wave into a light beam focused onto a moving spool of motion picture film. This method captured, with surprising exactitude, the voice's tiniest variations in pitch, timing, and volume as it delineated a melody, turning the signal into a wiggling horizontal line much like readings from our modern-day oscilloscopes. The technology had been used (by the aforementioned Gladys Lynch)* to study emotions in everyday speech, but Carl E. Seashore, a psychology professor at the University of Iowa, was the first to use it to analyze how the voice moves us so deeply when raised in song. Seashore and his colleague Milton Metfessel chose to study a style of singing renowned for its strong emotionality, but also considered untranscribable by standard musical notation: "negro folk singing," as they called it—that is, the work songs and spirituals sung by African American populations in Georgia, Tennessee, and North Carolina. Derived from the songs that slaves had created in the eighteenth and nineteenth centuries, this singing would go on to form the basis for all jazz, blues, and rock singing, including that of white artists like Elvis Presley, the Beatles, and Mick Jagger, to name only a few of the most famous.

The resulting study, *Phonophotography in Folk Music: American Negro Songs in New Notation* (1928), is a rare and remarkable document. In it, Seashore and Metfessel analyze the singing voices of African American farmhands and domestic workers, and they isolate a number of specific acoustic features that give the music its emotional power: including the bends in pitch that create the impression of sadness in blues singing, and something they call the *falsetto-twist*, which "resembles the voice breaking under emotion, especially grief."[3] But their primary discovery is how often voices that, to the naked ear, sound like they are perfectly on pitch, and strictly in rhythm, deviate widely from these norms, sliding well below or above the "correct" pitch, delaying an attack on a note for several milliseconds after the beat, or pouncing on a tone early, depending on the desired emotional effect. When Seashore and Metfessel, for comparison, analyzed the voices of trained opera singers, they professed

*See Chapter Three: Emotion

themselves "stunned" to discover that the most highly trained classical singers do the same thing as those folk singers: edging on or off a pitch, coming in slightly before the beat or slightly after. The authors concluded that such carefully modulated *imprecision* is a feature of all emotionally affecting vocal music—and, indeed, of all great art. The minuscule deviations from a mechanically "perfect" rendering of a song were akin to the expressive imprecision that the Renaissance Italians extolled as *sprezzatura* in oil painting and drawing, or that Chinese ink-and-brush painters prize in calligraphy and other painting, or Jack Kerouac promoted in the "automatic," off-the-top-of-his-head writing of *On the Road*—a freedom and brio that openly flouts perfection, but that gives vibrancy and emotional life. "The principle involved is a well-recognized theory of art," Seashore and Metfessel wrote.

This explains in part why so many of the current chart-topping songs in every genre (pop, country and western, hip-hop, rock), for all the virtuosity of their sometimes highly ornamented singing (glissandos, melisma, glass-shatteringly high pitches), sound cold, sterile, emotionally null: virtually all music producers now use Pro Tools software to "correct" the expressive slides on and off precise pitch. Many producers use the pitch-correction software to center the voice directly onto a note's mathematically determined frequency, and adjust the timing of every word so that the syllables fall, with metronomic exactitude, precisely on the beat. Robert Warsh, a singer friend of mine with an especially expressive and soulful voice, told me why the widespread use of Pro Tools has been so disastrous for so many genres of music. "It removes the human moment when the voice travels to the note it's seeking. I believe the listener registers, unconsciously, this natural 'pitch-finding'—and it's one of the things that attracts us to a voice. Compare a Sam Cooke song to one from Taylor Swift, and this will say it all."[4]

Swift provides a particularly salient example, perhaps, because her natural tendency toward a certain wobbliness of pitch is what gave her voice its charm and individuality in her early incarnation as an ingénue C&W singer. Her 2009 rendition on David Letterman's show of the superbly written "Fearless" (about a young woman plunging into her first romance) benefited from her obvious lack of vocal virtuosity, the slides on and off pitch, the inexpertly timed inhalations. Having since mutated into an arena-filling singer of machinelike, propulsive dance pop, Taylor's voice is today invariably autotuned and otherwise augmented, both in studio and live, to fit the robotically precise music. What is gained in sonic precision, immediacy, and commercial popularity is lost in the delicate emotionalism of her actual, appealingly imperfect, voice.

The Seashore and Metfessel study found that one property in particular epitomizes the artful "imprecision" that is the basis for all emotional expression in the voice: *vibrato*, a "flutter in pitch," a wavering in the fundamental frequency of the phonating vocal cords of a semitone interval (the difference between a white key on the piano and the black key adjacent), at a rate of five to seven oscillations per second, with a simultaneous alternation in volume between loud and soft.[5] That description sounds forbiddingly technical, but you actually know vibrato very well—you hear it every time you detect the quivering pulsation that singers like Ariana Grande or Beyoncé employ to give a plangent yearning to a long-drawn-out note, often at the end of a phrase. The brain converts the ambiguously pitched pair of notes into a single tone whose rapid movement between two frequencies mysteriously brings a lump to your throat or raises the hair on your arms.[6]

More than eighty years after Seashore and Metfessel's pioneering research, scientists are still debating the physiological origins, and psychological effects, of the vibrato. They have confirmed the authors' assertion that it occurs in all singing styles; you hear it in folk and pop, country and opera, hip-hop and death metal.[7] But it is used not only in singing:

every time you see and hear a classical violinist wiggle her finger on the fret board while bowing a note, or see a grimacing rock star perform the same finger-shaking gesture high up on the neck of his guitar to coax a few extra quivers of emotion from a solo, you are witnessing the vibrato. Trumpeters, clarinetists, trombonists, flautists use their lips in combination with the instrument's valves to create an identical fluttering of pitch and volume between two semitones—and at precisely the rate that Seashore and Metfessel first identified in the singing of African American folk singers of the 1920s: that is, between five to seven oscillations per second.

Seashore, in a later book exclusively about the vibrato,[8] called it "the most important of all the musical ornaments," and posited that its universality in music reflects its origins in the original musical instrument: the voice, where vibrato is a natural reflex of the vocal muscles when holding a sustained note. Great singers modulate its speed and duration through controlled movements of the laryngeal cartilages (shifting them back and forth like the guitarist's wiggling finger); rapid changes in respiration (pulsing movements of the diaphragm to vary the air pressure from the lungs and thus stress the periodic changes in volume from loud to soft that accompany the pitch changes); along with coordinated movements of the tongue, velum, and jaw that further shape the vibrato into its characteristic two-tone ululation. However, if the oscillation is too fast—that is, if it exceeds the seven oscillations per second that Seashore identified as the upper limit for an expressive vibrato—it creates an unpleasant effect that singers call "bleat"; too slow and you get the dreaded "wobble." Overextending the vibrato on a note, or using it on *every* note—troweling on an impasto of affected or mannered emotion—creates schmaltz. Some strict music aficionados insist that when the vibrato is executed properly a listener doesn't even consciously hear it, she merely responds with a powerful upwelling of emotion.[9] Why remains mysterious. Seashore thought the answer lay in the vibrato's evolutionary history. As he was the first to point out, all birds and nonhuman mammals produce the vibrato ("it is present . . . in the singing of the canary," he wrote, "in the bark of the dog"), and at the same rate as in humans, five to seven oscillations per second, in a semitone interval. It is likewise heard in our human laughter and sobbing.

No voice scientist, to my knowledge, has offered an explanation for the adaptive origins of this "phenomenon of sonance," as Seashore calls the vibrato.[10] I suspect the answer lies in Eugene Morton's observations (derived from Darwin's "principle of antithesis") about how all animals blend, at moments of indecision, the low pitch of aggression with the high pitch of submission. For the vibrato is, after all, a series of rapid up-and-down frequency changes across the held note. Seashore observed that vibrato does not evoke *particular* emotions—"we cannot distinguish feelings of love from hatred, attraction from repulsion, excitement from tranquility, by the vibrato," he wrote.[11] But perhaps this very ambiguity, like the "indecision" Morton detected in the animal voices he studied, is the basis of the "emotional instability" that Seashore said is the hallmark of the vibrato's universal, goose-bump-inducing power.

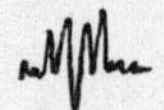

The emotional power of song is, of course, central to many forms of religious worship, from the chants of Shaman in indigenous tribes, to Islam's Call to Prayer (whose haunting strains I hear every Friday from the minaret of the mosque across the street), to the extraordinary "overtone singing" perfected by Buddhist monks in Mongolia: a hypnotic, unearthly style of vocalization in which the singer skillfully shapes the resonance chambers of throat, mouth, and lips to filter the vocal spectrum, accentuating certain overtones, muting others, so that particular pitches, ordinarily hidden within the rich array of harmonizing overtones, ring out with startling clarity, to create unearthly, piercingly high, whistling notes above the growling fundamental, like the zing of high tension wires in the wind. By filtering two overtones with differing frequencies that beat against each other, the most skilled throat singers can introduce a wowing rhythmic throb in the keening high tones, while simultaneously muting the growl of the phonating vocal cords. The effect (as this might suggest) is indescribable. Various other cultures and religions have developed forms of throat singing, including Tibetan monks and singers of a southern region of Siberia called Tuva, where the throat singers have incorporated the sounds into a religion of pastoral animism that sees all of nature, including animals, rocks and stones, running water and rushing wind, as infused with spirit, the uncanny whistling, groaning, growling tones produced by Tuvan shaman an evocation of the "voices" that emanate from all natural objects.

In Judaism and Christianity, the human voice is a metaphor for the soul. Because voice is produced by air pushed from the lungs and out through the lips, it necessarily evokes the original act by which God infused humans with life—literally "inspiring" us (from the Latin *inspirare*, "to blow into"), as described in Genesis 2:7: "And the LORD formed man *of* the dust of the ground, and breathed into his nostrils the breath of life." In Islam, life is also engendered by the inspired breath: the Koran calls it *al-Ruh* (which also translates as "soul"). Allah blows *al-Ruh* into a human when life begins, and *al-Ruh* leaves the body at death.[12] No wonder, then, that singing—extended musical *exhaling*—has propelled religious ceremonies for centuries, making of every "human being . . . a beautiful breathing instrument of music made by God after his own image," as Hyun-Ah Kim puts it in her study, *The Renaissance Ethics of Music: Singing, Contemplation and Musica Humana* (2015). The very act of singing, she says, becomes a "way of engaging the whole self in praising of God."[13]

In fact, the singing voice as an aid to religious worship in early Christianity gave rise to modern Western music as we know it, from classical to pop and everything in between. Jerry Lee Lewis called rock 'n' roll "the devil's music," but rock's origins, too, are in the plainchant singing of the Catholic liturgy, known as the Gregorian chant. Taking its name from the then-sitting pope, Gregory I (540–604), the Gregorian chant creates a mood of worshipfulness through an array of vocal strategies, including rules that strictly govern how melodies move. Dramatic leaps in pitch from low note to high, or vice versa, are rare: when chants mount upward in an ecstatic ascent toward heaven, they usually do so only one "step," or note, at a time; when jumps occur, they are generally of two or three tones, never more than five. Latin passages from the Bible supply the chants' "lyrics," which are always performed without instrumental accompaniment, to put the greatest possible emphasis on the voice, the text, the Word. When performed in groups (or choirs), the chants are always sung in strict unison without any distracting harmonies. The actual melodies, meanwhile, are restricted to a relatively narrow range (usually a single octave), which also lends a certain measured sobriety to the songs. But the Gregorian chant achieves its most hypnotic and unearthly effect from the absence of any regularly repeating rhythmical pattern, any artificial *beat*. This "unhooks" the chant from the internal metronome of the pulsing heart and blood, the time signature for everything else we do in life, from walking to talking, and places the chant in an ethereal realm that transcends time.

The emergence of the Gregorian chant marked the moment when church composers began to write down vocal melodies with a notation system that uses the familiar musical staff, with notes arranged in stepwise fashion—an exceedingly useful technology for a religion with expansionist ambitions. Because the all-important liturgical chanting no longer needed to be taught face-to-face in the oral tradition, it could now be widely disseminated, in the form of sheet music, with ease. And it was. While this helped to power the spread of Christianity around the globe, it also made the structure of vocal music "visible." Composers began to manipulate melodies and organize notes into specific groupings, which the church called "modes." These would evolve into the specific sequences of notes that we call scales, or keys.

In Gregorian chants, different modes give rise to different moods or emotions in listeners. Dom Joseph Gajard, a monk and musicologist who recorded Gregorian chants in the mid-twentieth century spoke of these emotions, variously, as "peaceful," "ecstatic," "refreshing," "enthusiastic."[14] The church borrowed ancient Greek terms to name the modes; the sad-sounding mode (a minor scale) they called the Aeolian; the happy, upbeat mode (a major scale) they called the Ionian. Over time, eight different "church modes" emerged, and each was given a (sometimes very daunting-sounding) name from the ancient Greek, like the Mixolydian, which is simply a scale (sing: *do re mi fa so la tee do*), with the *tee* flattened by a semitone. Songs written with the notes of this scale have an ambiguous mood, blending the "happy" major scale feel of the first six notes with a tug into inchoate sadness from that flattened seventh—as in John Lennon's "Norwegian Wood," which is in Mixolydian. Not that he knew it; Lennon couldn't read a note of music, but he knew how to evoke, through the grammar of song, the mixed excitement and despair of tumbling into an extramarital affair—or nearly doing so. Mixolydian's off-kilter mood also proved useful for evoking 1960s psychedelia, which is why Lennon used it for "She Said She Said" and "Tomorrow Never Knows."

The latter song, incidentally, has lyrics that Lennon lifted verbatim from a translation of the Tibetan Book of the Dead, a Buddhist text that describes what happens after death. Whether intentionally or not, the Beatles' arrangement of the song imitates Tibetan throat singing by setting up a single drone on C that carries through the whole song (like the low tone throat singers establish with the vocal cords) over the top of which Lennon sings, in his distinctively buzzing, "velar" voice,[15] a sinuous line of melody in Mixolydian ("*Turn off your mind, relax, and float downstream . . .*"), which mimics how the throat singers shape the resonance chambers to create an unearthly melody that soars above the hypnotic, meditative, timeless drone.

Science is learning what millions of people have known for millennia: singing heals spiritual malaise. In 2012, a team of investigators reported a significant decrease in anxiety and depression in cancer patients who were encouraged to sing,[16] and similar positive effects have been reported for people suffering from illnesses including arthritis and chronic pain. These positive effects are not simply psychological but are rooted in measurable physiological responses, including an increase in production of an immune system protein that fights infections in the upper respiratory tract.[17]

The most striking discoveries about the therapeutic effects of song involve choral singing. Even at an amateur level, the act of blending your voice with others in song causes the brain to secrete the chemical oxytocin, a hormone that creates the warm sensations of bonding, unity, and security that make us feel all cuddly toward our children and others we love—or infuses us with spiritual awe.[18] The hormone is found in all mammals, which suggests that it evolved as a way to encourage cooperation and group bonding. In early humans, group singing seems to have evolved as a tool for encouraging our natural sociality, for building loyalty, and for giving solace in moments of shared grief and pain.

This is how President Obama used it in that Charleston church service for the victims of the mass shooting. Valerie Jarrett, his senior advisor throughout his presidency, revealed in an appearance at the Aspen Institute in 2015[19] that she and Michelle Obama expressed skepticism when Obama told them, on the helicopter flight from Andrews Air Force Base to the memorial service in South Carolina, that he might sing. "Why on earth would that fit in?" Mrs. Obama asked, according to Jarrett, who had actually begged him *not* to sing on an earlier occasion: a fundraiser at New York's Apollo Theater in 2012. He had defied her by bringing down the house with a short snatch from Al Green's "Let's Stay Together." That had gone well, but Jarrett was no more enthusiastic than Mrs. Obama about the president singing at the church memorial. Obama, however, explained why he thought he should sing. He felt no special desire to perform a *solo*. "I think if I sing," he told them, "they might sing along with me." Intuitively, Obama knew what researchers had discovered about the healing oxytocin flood that comes with choral singing. And his gamble paid off. He was not three words into "Amazing Grace" when the entire congregation was on its feet, singing with him. For the first time during that service, they were smiling.

There is, I believe, a reason that Valerie Jarrett and the first lady discouraged President Obama from singing that day. Singing leaves us unusually naked and exposed. Think how easy it is to prattle away in conversation with your work colleagues—and how mortally embarrassed you would feel if required to sing "Happy Birthday" to them in a solo, a cappella performance. Even highly experienced performers feel the gulf between tuning the vocal cords to speech and tuning them to song. In 2017, the comedian and actor Bill Murray collaborated, as a singer, with cellist Jan Vogler, and as he told the *New York Times*, "something really different happens when you sing, it's not like talking or even telling a joke. When you sing, it's really, you are expressing yourself. It is a representation of yourself."[20] Murray doesn't explain why singing should be any more a "representation" of the self than speaking, but we know instantly what he means. By accentuating the rhythmic and melodic channel of the voice over that of the earthbound plod of articulate speech—by riding the exhaled breath through a sequence of pitches and beats that imposes on the air a pattern of vibration that we recognize as beautiful, healing, unifying, and emotionally nourishing—we not only cut to the quick of our humanity, but we reveal private dimensions of the self in ways that the cagey rhetoric of language can obscure. It is for this reason, I believe, that Jarrett and Mrs. Obama discouraged the president from daring to launch his solo voice into song: by doing so, he placed himself in a particular kind of peril, the peril that attends an unguarded opening of the heart in public view—not something that presidents (or their advisors) ordinarily expect the commander in chief of the world's largest military to do. Which might explain why Barack Obama was the only sitting president in history to risk it.

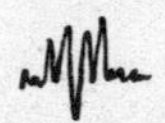

Even the most highly trained classical singers recognize that producing a musical note with the voice represents the purest and most naked self-revelation. In her 2004 memoir, *The Inner Voice*, the celebrated opera soprano Renée Fleming calls singing "an exercise in vulnerability"[21]—a vulnerability all the greater for divas, like Fleming, who deliberately place themselves under the scrutiny of critics and aficionados whose expertise in spotting flaws in technique or expression can be paralyzing.

Fleming's memoir offers the best account that I have been able to find of how a singer blends her natural, "untamed" gifts with the rigors of extensive training—while still maintaining and, indeed, even enhancing the emotional power of her singing. The daughter of singing teacher parents, Fleming grew up in an environment saturated by song ("Music was language in our house," she writes. "It was air"),[22] and she clearly inherited a lot from her parents: not only in the genetic lottery of her vocal instrument, the suppleness and health of her vocal cords, the size and shape of the resonating chambers of her chest and neck and head, but in the infectious passion her parents had for singing. Even before Fleming could speak, her mother was prompting her to "parrot back" sequences of notes sung to her—a feat the infant Fleming could perform with remarkable precision. In school, she landed the leading parts in musicals (at twelve, she played Eliza Doolittle in *My Fair Lady*), but as with any art form, shaping raw talent into a long and successful professional career is another story altogether, and Fleming writes, fascinatingly, of the technical mastery behind what she calls the soprano's "cultivated scream." It took her years to learn the physical and mental techniques that go into producing the impossibly loud and high notes she can make. Newborns do it instinctively—positioning the tongue and lips and larynx in the optimal arrangement for boosting certain overtones in the voice spectrum, to achieve house-shaking volume without excessive strain on the lungs or vocal cords.

In opera, the amplification of vowel overtones is called the "singer's formant," and Fleming learned to do it through creative visualization. She "imagines" that she is projecting her voice into highly specific targets in her body—"aiming sound mentally," as she puts it. For the highest notes, she targets the "mask"—the nose, cheekbones, and sinuses. Only then can she engage the involuntary muscles in the diaphragm, larynx, tongue, and face that allow her to project her voice "to the back of the hall without strain."[23] How she shapes that sound into something we deem "beautiful," so that each note hangs for a moment in the air, as present as an abstract Brancusi sculpture—shaped and shimmering in space, textured, polished, and conforming to all the criteria of proportion and harmony that Plato said embody perfection in the arts—well, that's another question entirely.

Science has been trying to penetrate that mystery and Fleming is unusual, as one of the world's most successful singers, in having lent herself to the effort. In 2017, she volunteered as a guinea pig for experiments into the neuroscience of singing conducted by the Kennedy Center and the National Institutes of Health. She spent two hours lying inside the narrow tube of an fMRI scanner, repeatedly singing one of the most emotionally resonant songs in her repertoire: "The Water Is Wide," a plangent Scottish folk ballad. All the expected areas of her brain "lit up" with activity: her Broca's and Wernicke's areas (as she produced the lyrics), her motor cortex (as it sent instructions to the larynx and articulators), her limbic structures (as they processed the song's emotions), and areas on the right side of the brain (that compute melody and rhythm). In short, the same structures activated in speech. Apparently, even the most high-tech wizardry is as yet powerless to explain why singing is so powerful *as singing.*

The most illuminating reflections on the power of singing that I have come across are from people who have spent their lives helping *others* achieve the fullest emotional expression with their voice. Laurie Antonioli is a singer, recording artist, singing coach, and chairperson of the Vocal Program at the California Jazz Conservatory. She believes the voices that move us the most have achieved a ruthless honesty of expression: they have been quenched of the mannerisms, affectations, trendy ornaments, and derivative stylistic tics that may make for massive pop hits, but whose emotional penetration is less than skin deep:

true auditory cheesecake (or "ear candy" in music biz parlance). This type of singing is especially frowned upon in the musical genre that Antonioli specializes in: jazz. Billie Holiday singing "Strange Fruit," Nina Simone scatting at the end of "Feeling Good," Joe Williams and Jon Hendricks singing "Going to Chicago" with Count Basie, Aretha Franklin singing *anything*—these are experiences at the furthest possible remove, aesthetically, emotionally, "spiritually," from an ear candy pop hit.

You don't have to be a music expert to know this. You can hear, and feel, the difference in emotional depth and power between the singing of any of the above-named artists and even as gifted a pop belter as Ariana Grande or a vocal gymnast like Beyoncé. But it is not always so easy to *explain* those differences. Antonioli calls it "authenticity"[24]—then instantly admits that "authenticity" can't be defined. "But everyone knows it when they hear it," she says. "I hold these workshops where people get up and sing to the group. People want to please listeners, to *sound good*, so they make the mistake of *manipulating* the voice. They'll try to sound seductive or sexy. Or powerful and commanding—whatever the song dictates. But when a singer gets to that authentic voice, when they get past the conscious manipulation, a totally different sound comes out and everyone in the class just gasps—they literally go *ahhhhh!* It's unmistakable. It has nothing to do with technical perfection. The voice could be shaky or soft, or whatever. It's the heart and soul."

The Self escaping into the Open.

The word "soul" comes up a lot when people talk about singing. Fleming uses it in her memoir. Antonioli used it. *Everyone* uses it. The soulfulness of singing is (as we've seen) at the basis of the art's central role in virtually all religious observance, back to stone age tribes and their chanting Shaman. Darwin, however, explained the deep emotions triggered by singing not as the stirring of a God-given soul, but as a physiological reaction that, in early humans, evolved to aid us in our survival and reproduction. In the *Descent of Man*, he said that the emotive content of all vocal expression—from singing, to poetry recitation, to public oratory—is an inheritance from our animal ancestors, and he even allows himself to wax poetic in describing how the stirring power of the human voice results from a kind of collective memory of ourselves in an earlier state of our evolutionary development: "The sensations and ideas thus excited in us by music, or expressed by the cadences of oratory, appear from their vagueness, yet depth, like mental reversions to the emotions and thoughts of a long-past age."[25] A few pages later, he expands on this: "The impassioned orator, bard, or musician, when with his varied tones and cadences he excites the strongest emotions in his hearers little suspects that he uses the same means by which his half-human ancestors long ago, aroused each other's ardent passions, during their courtship and rivalry."[26]

This is fine as far as it goes, but it doesn't go far enough. If Darwin correctly identifies the "animal" origins of singing's primal emotional appeal, he cannot account for its peculiarly *human* dimension, that part of singing that expresses our sense of belonging to a community of interdependent, empathetic, and cooperative human beings. For the congregation at that Charleston church, Obama's singing voice was more than an atavistic cry of pain or sorrow, or "mental reversion" to some more primitive expression of those inner states. When Aretha sings "Amazing Grace," or Paul McCartney sings "Hey Jude," or Pavarotti sings "Nessun Dorma," they shape the voice signal in very particular ways, to create a sense of beauty, to appeal to our higher aesthetic sense, to aid the heart's "ascent to Heaven," as the creators of Gregorian chant put it.

I am not religious in any conventional sense, but I would be less than honest if I didn't admit that, of any activity I have engaged in, singing is the one that came closest to convincing me of the existence of the "soul"—if only because I lack any better term for whatever very private, but vital and *essential*, part of myself I seemed to be accessing, and releasing into the air, through the act of singing. When thinking about why projecting my voice into a melody should have been so pleasurable, cathartic, aesthetically pleasing, rewarding, and restorative, I am thrown back on what Julie Andrews said about the "ecstasy" of singing over a full orchestra, or what the middle-aged opera tenor said to me before undergoing microsurgery with Dr. Zeitels to remove the scarring on his vocal cords: "I've grown a little tired of just *talking* about it. I mean, when you sing, you're giving voice to your *soul*." I'm not sure I ever would have made such grand-sounding claims for my own singing, back when I could actually do it—before I injured my voice with Jann's band. Now, I feel differently. As the 1980s hair metal band Cinderella (and Joni Mitchell, in a very different context) sang: you "don't know what you got till it's gone."

CODA

The truth, of course, is that we all lose our voice, sooner or later. Like everything else in nature, it has a life cycle. After its birth in a scream and the big hormone-driven changes of puberty, the voice settles into a long period of stability that can last forty, fifty years. But ultimately, it gives over to old age. The time of onset is different for every individual, depending on genetic and environmental factors, but it arrives eventually, the result of a cascade of changes throughout the body as it ages. In men, reduced testosterone production causes the vocal cords to thin and shrink, making them more like those of women, which is why comedians, in imitating crotchety old men, often speak in a scratchy falsetto. Women, meanwhile, undergo their own hormonal changes later in life, with menopause, a flood of extra estrogen that actually causes the vocal cords to swell, lowering the vocal pitch, bringing it closer to the male voice. In short, with the passing of sexual potency and reproductive power, the dimorphism of the human voice, so crucial to the propagation of the species, vanishes. Men and women grow to sound the same.

The voice also takes on a creaking, rickety sound that is the result of what voice specialists call "presbyphonia," the natural deterioration and decay of the vocal cord tissue itself. The muscles that run through the vocal cords stiffen like the hamstrings on an aging runner, the soft collagen layer over top of the muscle breaks down, the protective layer of mucus membrane hardens—so that the vocal cords no longer perform the sinuous, liquid rippling motion characteristic of their vibration in young people, and that lends healthy youthful voices their rich complexity, depth, and nuance. Aged voices grow brittle-sounding. Couple this with the natural weakening of the muscles that move the laryngeal cartilages, to say nothing of the deterioration of the nerves that signal to those muscles, and you get voices that are increasingly robbed of prosody, stuck in a perpetual emphatic shout (worsened by the deterioration of the speaker's own hearing), or a mumbling half whisper. Professional singers who have spent their lives effortlessly hitting the pitches they want suddenly find it difficult in their late sixties (or before) to stay in tune. Meanwhile, the speed and clarity of the speaking and singing voice vanishes as the articulatory muscles of the face and tongue weaken. Speech slows, grows blurry and filled with pauses.

The decrease in volume and power of the voice in the elderly is thanks to the deterioration of other muscles throughout the body including the diaphragm and the intercostal muscles between each rib. The torso itself shrinks, the lungs become smaller, stiffer, less pliable. With a weakened bellows, the voice emerges increasingly as a whisper, a condition made more acute by the tendency of older people, owing to the curvature of the spine, to lean forward, further constricting the action of the lungs.

Environmental factors aggravate the voice's decline. Smoking gives rise to emphysema and other respiratory ailments. Drinking bathes the delicate vocal cords in abrasive alcohol and promotes reflux disease, when scouring acids from the stomach travel up the esophagus and scorch the vocal membranes. The vocal tract itself, meanwhile, changes, altering the voice's resonance, as the nose falls backward into the face, reducing the size of the nasal resonators and as fat collects in the neck, squeezing the resonance chamber of the throat, changing the voice's overtone structure, and thus timbre, to the point where even loved ones might not, over the phone, recognize you anymore.

Science and medicine can do only so much to arrest the voice's natural decline. For a biological function dependent on so many diverse parts of the body, the war against the aging voice has to be fought on many fronts simultaneously. Aerobic exercise and strength training can arrest some of the diminishment related to the muscular atrophy that reduces lung power; changes in diet can stem some of the deterioration associated with acid reflux. But by far the stiffening and thinning and atrophying of the vocal tissue is what makes us sound old. Which is why, when I interviewed Dr. Zeitels back in 2012, he said that a therapy for restoring the youthful pliability of the vocal cords was the Holy Grail of voice science.[1] Inspired by cosmetic surgeons who transfer fat from the buttocks into areas of the face to plump out wrinkles, some surgeons had tried injecting fat cells into the vocal cords in a bid to restore their natural pliability. But in rare cases, the fat turns to scar tissue—making things worse. In his quest for an "artificial vocal cord," Zeitels was developing an injectable biogel, a water-based polymer that, when sitting in a lump on a work bench in his lab, could very closely reproduce the liquid ripple of an actual vocal cord, and which Zeitels hoped could be injected under the thin layer of mucus membrane to replace the vibratory layers that gradually stiffen and wither with age. Zeitels had actually come up with the idea when treating Julie Andrews after her disastrous surgery at Mount Sinai, in the late 1990s, as a way to put back the tissue that had accidentally been lost. At the time of our interviews, Zeitels had not yet put the therapy into human trials, but he was already dreaming big about the gel as a vocal Fountain of Youth—a therapy that would make eighty-year-olds sound like they are young again, and which would make the aging rockers whom Zeitels treated, like The Who's Roger Daltrey and Aerosmith's Steven Tyler, sound as they did in their thirties. Tyler and Daltrey, then in their fifties, were experiencing a kind of accelerated aging of the voice through the sheer wear-and-tear of their decades-long careers. Teachers and lecturers and others whose jobs require heavy use of the voice experience a similar deterioration. "It's not senescent tissue," Zeitels told me, "it's tissue that has been *hit* more times," as the vocal cords slam together trillions of times over dozens of years, in phonation. Zeitels was boundlessly optimistic about the powers of medical science to turn back the clock on the voice. "It's not 'Is it going to happen?'" he told me. "It's *when* it's going to happen." Perhaps. But almost a decade later, it hasn't happened yet.

When I learned about the psychosocial factors associated with the aging voice—the withdrawal from social life, the self-recusal from formerly enjoyable activities like singing in church, or joining family and friends at too-noisy restaurants—and as I read about how these actions increase loneliness and give rise to depression and poorer overall health, I was brought up short. This cascade of symptoms in the elderly is a little too close to the changes I noted in my own life, after age *forty*, because of my vocal polyp. Which perhaps explains why, in my midfifties, I threw all caution to the wind and, when invited by a fellow *New Yorker* staff writer, John Seabrook, to join his amateur rock band, the Sequoias, I jumped at the chance (just as I had with Jann all those years before). At the time, the Sequoias were a motley assemblage that included *New York* magazine senior editor John Homans, on lead guitar, and an inevitable rotating cast of bass players (NPR producer Charlie Foster, musician-ringer Rowlie Stebbins) and drummers. For the latter, we finally settled on *Elle* senior editor Ben Dickinson who also, fortunately, possesses a high, clear singing voice that made possible our covering songs by the young Paul McCartney, like "I Saw Her Standing There." Otherwise, musical ability was in short supply. The chief criterion for membership in the group was being taller than six feet (hence, the band's name). I played keyboards and was, at first, determined not to sing, since I knew that this could only further damage my voice. But soon enough, I found myself grabbing the mic and wailing away with my usual incautiousness—and why not? If all of our voices are scheduled, eventually, to be silenced, why not *go out singing*?

This "swan song" for my voice proved to be a bigger deal than any of us could have anticipated when, in the final year of Barack Obama's presidency, *New Yorker* editor David Remnick (six-foot-two) joined the Sequoias as a guitarist. ("Sequoias?" he quipped, upon being told the band's name. "As in, tall, old, and dead inside?") The glow of his fame and power led to an invitation to play at the White House Correspondents' Dinner—not the main event itself, I hasten to add, but one of the many scheduled parties and jamborees that took place across the capital during those weekends: in our case, an annual party held in a ballroom at the Fairmont Hotel, and run by the Rolling Stones' touring keyboardist, Chuck Leavell, and his environmental charity, the Mother Nature Network.[2] Like every boomer-heavy Dad Band, we played the usual collection of Beatles and Stones numbers, but also the Clash cover "Brand New Cadillac," which featured me on lead vocals, if what I did onstage that night can be dignified with the term. My voice sounded like sheet metal being torn by some massive machine, all roar and growl and shearing overtones, doubtful in its pitch, one-dimensional in its dynamics (a gasping shout), and I messed up some lyrics. I was hoarse and growling for weeks afterward and I am not at all sure that my time with the Sequoias did not further damage what remains of my voice. But I don't regret a thing.

Nor do I any longer contemplate having my vocal polyp removed. I did think about it in the immediate wake of writing my story about Dr. Zeitels. And I thought about it again when I signed on to write this book. I believed it might make a nice ending—my Lazarus-like rebirth as a singer and occasional public speaker. But I ultimately declined to go under the knife. Which might seem nonsensical, given what working on this book has taught me about the incalculable importance of the voice to us as individuals and as a species. In defense of my decision, I can offer only a paraphrase of the great fashion designer Coco Chanel, who said that "at fifty, you get the face you deserve." I turned sixty-one during the writing of this book. And my voice, with its nicks and scars and telltale rasp, tells its own history of my life, just like yours does.

Earlier, I said that this book serves no self-help function, does not purport to dispense tips on how to achieve a more powerful, or sincere, or persuasive voice. But in these closing remarks, I find myself reflecting on the fact that the most effective and expressive voices—those that *connect* with listeners in ways that change behavior or habits or ideas—are those that form the most direct channel between the speaker's interior life and the sounds that emerge from the mouth. Language is only one layer of that complex acoustic signal and, arguably, not the most important one. Language that does not sing, that does not acknowledge the stirring melody and movement of thought, that is not animated by the dancelike syncopations of rhythm, of linguistic and emotional prosody, is dead language. The finest of Shakespearean sonnets cannot move a listener when spoken in dull monotone. Which is why the most emotionally rousing and intellectually stimulating speech, whether dispensed in private conversation or at the podium, revels in the full orchestration of the voice, from the bellows of the lungs, to the fluid pitch shifts of the vocal cords, to the rhythmic and percussive articulations of the tongue and lips. Speech and song are equally an assertion of our existence against the void, a means for animating the air with news of our presence, however ephemeral, and thus should be performed with confidence in the Self, and with an awareness of the music from which our linguistic capability arose. So project your voice without fear or favor, weaponize it should the legitimate need arise, soften it when the mood calls for it, but be aware of its full, fantastic range of expression, and revel in it. That, in any case, is what I intend to do with what remains of my scarred and aging voice. The voice I deserve.

ACKNOWLEDGMENTS

In this book's Introduction, I said that it was "suggested" to me, by a reader of my Dr. Zeitels article, that there might be a book on the subject of the voice. Here, in the less formal confines of the Acknowledgments, I feel comfortable divulging that the suggestion came from Jonathan Karp, the President and CEO of Simon & Schuster, when we were blue-skying book ideas in his office. So my first thanks are owed to Jonathan—and also one of his senior editors, Karyn Marcus, whom he called in to pow-wow about the idea. That I overcame my initial trepidation and doubt about actually attempting such a book ("It's simultaneously too *narrow* and too *wide* a subject!" as I was given to telling anyone who would listen) is thanks to my agent of thirty years, Lisa Bankoff, who pushed me to *finish the proposal*.

Because I never really stopped feeling daunted by the book's ever-widening scope, even after inking the book deal and starting to write, my loudest hosannas of thanks and praise are for this book's editor, Eamon Dolan—who also happened to be the first to acquire it (when he was working for a completely different publisher: long story). That Eamon is himself a singer and (uncannily) also performed Yeats's "Second Coming" in a high school public-speaking competition in part explains why he, too, "heard" a book on the subject of the voice. But it's what he did to make that book actually come into existence that I feel compelled to celebrate here. It is difficult to describe what makes a great editor, but for this writer (at least) it has something to do with the almost mystical way that they impart confidence, how they make you feel as if you are (or can become) *the* authority on a given subject. I won't bore the reader (or embarrass myself) by describing the condition of the early drafts that I showed to Eamon, but let's just say that tact, patience, supreme confidence, and nerves of steel are also key attributes of a great editor—how else to explain his not actually yelling at me that I was *insane* to imagine that a reader would wade through a detailed discourse on auditory physics in the first chapter? If only I could claim that this was the sole instance when I sailed serenely into the weeds. But Eamon's skills are by no means limited to hacking a clear path through trees that have utterly obscured the forest. For instance, he suggested that the book could bear a whole section on the Lincoln-Douglas debates. As a Canadian, I had (I'm embarrassed to admit) limited knowledge of the debates—and besides, *a whole section*? Well, suffice to say that when I began to dig into the history, I was thrilled to see how beautifully that episode of one-on-one vocal combat illuminated the larger theme that had been taking shape under my pen: the centrality of the human voice in the unending struggle between freedom and enslavement. This addition, along with Eamon's cogent suggestions about the singing voice, oratory and rhetoric, the voice and religion, to say nothing of a big breathtaking structural change that he floated as a "modest proposal" late in the game, made this book not only immeasurably better, but actually (I hope) readable. Meanwhile, any slow, overly wonky sections that persist (the ones you skimmed or skipped altogether) represent moments when, out of ego or stupidity or stubbornness, I defied Eamon's constant reminder that "You're not writing this for scientists."

I did, however, rely on a lot of scientists. Philip Lieberman granted me two long phone interviews that proved indispensable, but that did not lend themselves to direct quotation. The same goes for conversations I had with Johan Sundberg, Krzysztof Izdebski, William Labov, Ingo Titze, Branka Zei-Pollermann, as well as some nonscience authorities, including Audrey Morrissey (the executive producer of the hit TV show *The Voice*) and Rachelle Fleming (sister to Renée and herself a gifted singer). Klaus Scherer, Björn Schuller, John Baugh, and John McWhorter were all very generous with their time, as was my neighbor Andrea Haring, whose brain I ruthlessly picked every time we ran into each other in the laundry room or elevator. Her boss, the legendary Kristin Linklater, who died in June 2020 at age eighty-four as I was writing these closing remarks, also graciously shared thoughts on the voice with me over email from her redoubt in Orkney, Scotland. Daniel Everett, field linguist extraordinaire, and my host in the Amazon, kindly sent me the relevant chapters on the evolution of the human voice from his then book-in-progress, *How Language Began* (2017).

One of the considerable challenges in writing about the voice is that scientific research on the subject fell off precipitately in the late 1950s when Noam Chomsky convinced the scientific world at large that all the significant action regarding human vocal communication took place up in the highest levels of the brain and that the voice was just an "epiphenomenon"—a secondary consideration, at best. Thus, some of the most salient pioneering research on the voice was relegated to the scrap heap of science history and can, today, only be found in the rare books department of the New York Public Library—and sometimes not even there. Walter Rudolf Hess's seminal work on the roots of mammalian emotional vocalization in cats was impossible to lay my hands on anywhere until I remembered the rare book collection at the New York Academy of Medicine Library at 103rd Street and Fifth Avenue (a refuge I had discovered back in 2005 when writing about medicinal leeches for *The New Yorker*). I offer hearty thanks to the collection's head librarian, Arlene Shaner, who dug up for me piles of criminally forgotten research on the human voice from the first half of the twentieth century: books, monographs, and journal articles—including a yellowing first edition of Negus's pioneering 1929 tome, *The Mechanism of the Larynx.*

One book I could not find *anywhere* was written by Philip Lieberman's coinvestigator into the Neanderthal voice: Edmund Crelin, the Yale anatomist who did groundbreaking work on the elevated larynx in newborns. In 1987, Crelin published *The Human Vocal Tract: Anatomy, Function, Development and Evolution,* a book that expanded on his work with Lieberman and that included photos of the macabre simulated Neanderthal throat, tongue, and mouth that Crelin constructed from silicone and wire to learn how our extinct human relatives sounded (by using a toy noisemaker for phonation). I was determined to see that elusive volume. Crelin died in 2004, but an email to one of his daughters, in Florida, resulted in her sending me a pristine hardcover copy of this rare and fascinating book. Thank you, Sherry Crelin! And thank you, too, to the staff and librarians at the New York Society Library, where I wrote the bulk of this book before the COVID-19 quarantine shut it down as I was writing the final chapter.

It would not be possible for me to remember, let alone thank, all the many people to whom I spoke in casual conversation about this project and who, in response, offered their own spontaneous reflections on the voice and thus helped to shape my view, or sent me down yet another path of investigation. But three people should be mentioned for the perspicuity of their reflections: my friend, author Bill "Chip" Doyle, who made the funny and honest comment about the dreadful *insincerity* he detected when hearing his taped voice for the first time (I should add that Chip is anything but insincere—which is what gave his comment its special salience); my friend Charlotte Harvey, an actress and artist who actually attended one of Kristin Linklater's "Freeing the Natural Voice" workshops in Scotland in 2017, and who supplied the detail that every attendee, at some point, cried about his or her mother. During this same conversation, Charlotte surprised me by asking, "How deep are you going—I mean, are you going to the molecular level? The FOXP2 gene?" This was during one of the many times when I was struggling to find the proper focal length for the book. How deep was *too deep*? I knew that FOXP2 was central to my argument that our voice made us human and propelled us to the top of the food chain, but I was still trying to decide if descending to the microscopic level of minuscule amino acid substitutions on a single DNA strand was TMI, as the kids would say. Charlotte gave me timely confirmation that it was not. The third "layperson" I must single out for special thanks is New York City doorman Big Ray Hernandez, who comes from a long line of professional wrestlers (and who also moonlights as a wrestling podcaster and occasional voice-over artist). It was Ray, to whom I was jaw-boning about the book in my building's lobby, who told me about the relevance of Donald Trump's years as host of Vince McMahon's WWE wrestling championships to his election as President—comments that sent me down an avenue of investigation that I might otherwise never have pursued.

At a time when the physical (as opposed to virtual) book is so much to be cherished for its increasing rarity, I thank everyone at Simon & Schuster who made this such a beautiful physical object to hold and behold, from the superb cover designer, Jim Tierney, to the person in charge of its page design, Ruth Lee-Mui (to whom I owe added thanks for having accepted my suggestion that to mark line breaks, we use a sound wave sketched by my artist/illustrator wife, Donna.) Fred Chase executed the copyedit, a term that in no way does justice to what he brought to this book. Correcting grammatical slips and bringing a book into conformity with a publisher's house style is crucial, but Fred also happens to have performed a fact check throughout, eliminating many errors, and also bringing to bear his considerable knowledge of U.S. history and many other subjects that improved this book dramatically. For the paperback, I refined the Gregorian Chant section with the help of Patricia P.E. Janssen, an early music performer, researcher, and author. Any lingering mistakes, alas, are on me. For all logistical help, I offer a special salute to Eamon's unfailingly cheerful and helpful assistant, Tzipora Baitch.

There remain only a few friends and family to thank. Writer, editor, and musician Mark Rozzo was a great sounding board at all parts of this process. Likewise, writer Joe Hooper and ex-senior editor at *Elle* Ben Dickinson (drummer and singer for the Sequoias), who caught a few errors that would have truly made me look asinine. Thanks to my friend Chris Deri, who happened to have begun writing his first novel as I was embarking on this book and who was thus always eager to compare notes on how agonizing writing can be, and who listened with amazing patience as I detailed, for the millionth time, why I didn't want to begin the book with a description of my vocal injury ("The memoir-styled beginning has become such a cliché"—a bit of misguided dogma from which Eamon saved me when he said, one day, "You know, John, I think we need to hear a little *something* about your own voice, higher up."). Special thanks go to my son, Johnny, whose life I occasionally ransacked for examples of voice and speech, and whose response, when I told him about the "evolution" angle to this book, was "Hey, cool!" (As a then-college freshman, he'd been reading *Sapiens.* Sometimes an eighteen-year old's endorsement is all you need.) Finally, I extend my deepest thanks, and inevitably some apologies, to my wife of thirty years, Donna Mehalko, to whom I will often read aloud a difficult passage or chapter, since I know that any falseness, grandiosity, or evasive "fanciness" that I have introduced into the prose will manifest itself in a fatal, faltering, tinny note in my voice that is instantly audible to her (and thus to me). At which point, it's back to the drawing board.

—JC, June 13, 2020

NOTES

INTRODUCTION: PERSONALLY SPEAKING

1. John Colapinto, "Giving Voice," *The New Yorker,* February 24, 2013.
2. Johan Sundberg, *The Science of the Singing Voice* (Dekalb: Illinois University Press, 1987), 2.
3. Aristotle, *De Anima* (London: Cambridge University Press, 1907), 89.
4. Paralanguage was first described and named by linguist George L. Trager (1906–1992) while he was working for a unit of the US State Department that taught American diplomats how to comport themselves abroad. Focusing on all vocal sounds "not having the structure of language," Trager created a complicated notation system that represented hundreds of noises ("moaning and groaning, whining and breaking, belching and yawning"). He first published his findings in: George L. Trager, "Paralanguage: A First Approximation," *Studies in Linguistics* 13, nos. 1 and 2 (1958): 1–12. He then turned to psychiatry for clues to what these noises might *mean*. Robert E. Pittenger, a psychiatrist at the State University of New York, was eager to marry paralinguistic and psychoanalytic methods and coauthored, with linguists trained in Trager's method, an ambitious "microanalysis" of a psychotherapy session between a housewife patient and her therapist, the first (and last) book-length paralinguistic study of its kind. Robert E. Pittenger, Charles F. Hockett, John J. Danehy, *The First Five Minutes* (Ithaca, NY: Paul Martineau, 1960).
5. Certain monkeys, meerkats, squirrels, chickens, and other avian species emit highly differentiated alarm calls to alert others of specific threats—one cry, for instance, prompts the animals to take shelter in bushes from an aerial attack like

that of a hawk, another triggers evasive action from a ground predator like a snake. But these sounds do not name specific predators, nor are the calls learned from parents, as words are. Instead, they are hardwired reflexes. For this reason, evolutionary biologist Tecumseh Fitch rejects the notion that these calls are precursors to language: "They're more like laughter and crying, which are also calls that are innate. You don't hear your mother crying to learn how to cry." Fitch quoted by Christine Keneally, *The First Word* (New York: Penguin, 2007), 115.

6. Yuval Noah Hariri, *Sapiens: A Brief History of Humankind* (New York: HarperCollins, 2015).
7. Author interview with Branka Zei-Pollermann, July 5, 2017.
8. Helen Blank, Alfred Anwander, Katharina von Kriegstein, "Direct Structural Connections Between Voice- and Face-Recognition Areas," *The Journal of Neuroscience* 31, no. 36 (2011): 12906–15, https://www.ncbi.nlm.nih.gov/pubmed/21900569.
9. Daniel J. Levitin, *This Is Your Brain on Music: The Science of a Human Obsession* (New York: Dutton, 2016), 138–39.

ONE: BABY TALK

1. A. Pieper, "Sinnesempfindungen des kindes vor seiner geburt," *Monatsschrift Fur Kinde-rheilkunde* 29 (1925): 236–41. Cited in Barbara Kisilevsky and J. A. Low, "Human Fetal Behavior: 100 Years of Study," *Developmental Review* 18 (1998): 11.
2. Denis Querleu et al., "Fetal Hearing," *European Journal of Obstetrics & Gynecology and Reproductive Biology* 29 (1988): 191–212.
3. William P. Fifer and Chris M. Moon, "The Effects of Fetal Experience with Sound," in *Fetal Development: A Psychobiological Perspective*, ed. Jean-Pierre Lecanuet (Hillsdale, NJ: Lawrence Erlbaum Associates, 1995), 351–66.
4. Anne Karpf, *The Human Voice* (New York: Bloomsbury, 2007), 81. Karpf cites a paper by Susan Milmoe et al., "The Mother's Voice: Postdictor of Aspects of Her Baby's Behaviour," Proceedings of 76th Conference of the American Psychological Association, 1968. Karpf also cites Suzanne Maiello, "Prenatal Trauma and Autism," *Journal of Child Psychotherapy* 27, no. 2 (2001).
5. "Infants less than two hours old react and orient more to the mother's voice than to those of other women." Melanie J. Spence and Anthony J. DeCasper, "Prenatal Experience with Low-Frequency Maternal-Voice Sounds Influence Neonatal Perception of Maternal Voice Samples," *Infant Behavior and Development* 10, no. 2 (April–June 1987): 133–42.

6. Anthony DeCasper and Melanie J. Spence, "Prenatal Maternal Speech Influences Newborn's Perception of Speech Sounds," *Infant Behavior and Development* 9, no. 2 (1986): 133–50.
7. Anthony DeCasper and Phyllis A. Prescott, "Human Newborns' Perception of Male Voices: Preference, Discrimination and Reinforcing Value," *Developmental Psychobiology* 17, no. 5 (1984): 481–91. Also: Cynthia Ward and Robin Cooper, "A Lack of Evidence in 4-Month-Old Human Infants for Paternal Voice Preference," *Developmental Psychobiology* 35, no. 1 (1999): 49–59.
8. Peter D. Eimas, Einar R. Siqueland, Peter Jusczyk, and James Vigorito, "Speech Perception in Infants," *Science* 171, no. 3968 (January 22, 1971): 303–6.
9. Peter D. Eimas, "Auditory and Phonetic Coding of the Cues for Speech: Discrimination of the /r-l/ Distinction by Young Infants," *Perception and Psychophysics* 18 (1975): 341–47; and Lynn A. Streeter, "Language Perception of 2-Month-Old Infants Shows Effects of Both Innate Mechanisms and Experience," *Nature* 259 (1976): 39–41.
10. Patricia Kuhl, "The Linguistic Genius of Babies," TED Talk, https://www.youtube.com/watch?v=G2XBIkHW954, uploaded February 18, 2011.
11. Paula Tallal, "Language Comprehension in Language-Learning Impaired Children Improved with Acoustically Modified Speech," *Science* 271, no. 5245 (January 5, 1996): 81–84.
12. Anne Cutler and Sally Butterfield, "Rhythmic Cues to Speech Segmentation: Evidence from Juncture Misperception," *Journal of Memory and Language* 31, no. 2 (1992): 218–36.
13. Patricia Kuhl, *The Scientist in the Crib: What Early Learning Tells Us About the Mind* (New York: William Morrow, 2000), 110.
14. Charles A. Ferguson, "Baby Talk in Six Languages," *American Anthropologist, New Series* 66, no. 6, part 2 (1964): 103–14.
15. Noam Chomsky, *Syntatic Structures* (The Hague: Mouton, 1957), 116.
16. Noam Chomsky, "Things No Amount of Learning Can Teach," *Omni* 6, no. 11 (1983), https://chomsky.info/198311__/.
17. This account of how Catherine Snow entered the field of language acquisition is from Snow's account in *Current Contents*' "Citation Classics," no. 1 (January 1985): 18.
18. Catherine E. Snow, "Mothers' Speech to Children Learning Language," *Child Development* 43, no. 2 (June 1972): 549–65.
19. Olga Garnica, "Some Prosodic and Paralinguistic Features of Speech to Young

Children," in *Talking to Children: Language Input and Acquisition*, eds. C. E. Snow and C. A. Ferguson (Cambridge: Cambridge University Press, 1977).

20. Anne Fernald, "Four-Month-Old Infants Prefer to Listen to Motherese," *Infant Behavior and Development* 8 (1985): 181–95.
21. Hojin I. Kim and Scott P. Johnson, "Infant Perception," *Encyclopaedia Britannica* online, https://www.britannica.com/topic/infant-perception.
22. Peter F. Ostwald, "The Sounds of Infancy," *Developmental Medicine and Child Neurology* 14 (1972): 350–61.
23. Roberta M. Golinkoff and Kathy Hirsh-Pasek, *How Babies Talk* (New York: Plume, 2000), 20.
24. Donald H. Owings and Debra M. Zeifman, "Human Infant Crying as an Animal Communication System," in *Evolution of Communication Systems: A Comparative Approach*, eds. D. K. Oller and U. Griebel (Cambridge: MIT Press, 2004), 160.
25. Peter F. Ostwald, "The Sounds of Emotional Disturbance," *Archives of General Psychiatry* 5 (1961): 587–92.
26. Arthur Janov, *The Primal Scream* (Venice, CA: Dr. Arthur Janov's Primal Center, 1999), 9–11.
27. Janov, *The Primal Scream*, 55.
28. Conversation with Charlotte Harvey, December 31, 2018.
29. Donald Eisner, *The Death of Psychotherapy: From Freud to Alien Abductions* (New York: Praeger, 2000), 51–52.
30. Philip Norman, *John Lennon: The Life* (New York: Ecco, 2009).
31. Peter Doggett, *You Never Give Me Your Money: The Beatles After the Breakup* (London: The Bodley Head, 2009), 220.
32. "John Lennon Talks About 'Mother' and Primal Scream Therapy, 1970," YouTube, https://www.youtube.com/watch?v=J5irvO7vzx8.
33. Birgit Mampe, Angela D. Friederici, Anne Christophe, and Kathleen Wermke, "Newborns' Cry Melody Is Shaped by Their Native Language," *Current Biology* 19 (December 15, 2009): 1994–97.
34. Owings and Zeifman, "Human Infant Crying as an Animal Communication System," 10.
35. Mary Carmichael, "Health Matters: Making Medical Decisions for Kids," *Newsweek,* January 30, 2009, https://www.newsweek.com/health-matters-making-medical-decisions-kids-77773.
36. Daniel Lieberman, Robert McCarthy and Jeffrey Bruce Palmer, "Ontogeny

of Hyoid and Larynx Descent in Humans," *Archives of Oral Biology* 46 (2001): 117–28.

37. James Booth et al., "The Role of the Basal Ganglia and Cerebellum in Language Processing," *Brain Research* 1133, no. 1 (2007): 136–44; and Steven Pinker, *The Language Instinct* (New York: Harper Perennial Modern Classics, 2007), 269.
38. Soo-Eun Chang and Frank H. Guenther, "Involvement of the Cortico-Basal Ganglia-Thalamacortical Loop in Developmental Stuttering," *Frontiers in Psychology*, 10 (January 28, 2007), 3088.
39. John Updike, *Self-Consciousness* (New York: Alfred A. Knopf: 1989), 79-111.
40. Katharine Davis, "VOT Development in Hindi and in English," *Journal of the Acoustical Society of America* 87 (1990), posted August 13, 2005, https://asa.scitation.org/doi/10.1121/1.2027880.
41. Eric H. Lenneberg, *Biological Foundations of Language* (New York: John Wiley & Sons, 1967), 146.
42. Lenneberg, *Biological Foundations of Language*, 150.
43. Lenneberg, *Biological Foundations of Language*, 158.
44. Russ Rymer, *Genie: A Scientific Tragedy* (New York: Harper Perennial, 1994), 90.
45. James T. Lamiell, "Some Philosophical and Historical Considerations Relevant to William Stern's Contributions to Developmental Psychology," *Journal of Psychology* (2009), 217, 66–72.
46. Charles Darwin, "A Biographical Sketch of an Infant," *Mind* 2 (1877): 285–94.
47. "Dad Has Full Convo with His Baby," YouTube, uploaded June 5, 2019, https://www.youtube.com/watch?v=0IaNR8YGdow.
48. Lynne Murray and Colwyn Trevarthen, "The Infant's Role in Mother–Infant Communications," *Journal of Child Language* 13, no. 1 (1986): 15–29.
49. Harvey Sacks, Emanuel Schegloff, and Gail Jefferson, "A Simplest Systematics for the Organization of Turn-Taking for Conversation," *Language* 50, no. 450part 1 (December 1974): 696–735.
50. A. M. Liberman, F. S. Cooper, D. P. Shankweiler, and M. Studdert-Kennedy, "Perception of the Speech Code," *Psychological Review* 74, no. 6 (1967): 431–61.
51. Liberman, Cooper, Shankweiler, and Studdert-Kennedy, "Perception of the Speech Code," *Psychological Review* 74, no. 6 (1967): 431–61.
52. N. J. Enfield, *How We Talk: The Inner Workings of Conversation* (New York: Basic Books, 2017), 42–43.
53. Enfield, *How We Talk*, 51–55, recounts experiments by Sara Bögels and F.

Torreira, "Listeners Use Intonational Phrase Boundaries to Project Turn Ends in Spoken Interaction," *Journal of Phonetics* 52 (2015): 46–57.

54. David Brazil, *The Communicative Value of Intonation in English* (Cambridge: Cambridge University Press, 1985).
55. Anne Wennerstrom, *The Music of Everyday Speech: Prosody and Discourse Analysis* (New York: Oxford University Press, 2001), 261.
56. Chad Spiegel and Justin Halberda, "Rapid Fast-Mapping Abilities in 2-Year-Olds," *Journal of Experimental Child Psychology* 109 (2011): 132–40.
57. Pinker, *The Language Instinct*, 270.
58. The sociolinguist Robbins Burling, in his essay "The Slow Growth of Language in Children," writes that "Producing a syntactic construction should be looked upon as only the final stage in a long developmental process," http://www-personal.umich.edu/~rburling/Slowgrowth.html.
59. Pinker, *The Language Instinct*, 29.
60. Matt Ridley, *Genome* (New York: Harper Perennial, 2006), 93.
61. Daniel J. Levitin, *This Is Your Brain on Music: The Science of a Human Obsession* (New York: Dutton, 2006), 191.
62. Roberta Michnick Golinkoff and Kathy Hirsh-Pasek, *How Babies Talk*, 50.

TWO: ORIGINS

1. Charles Darwin, *The Origin of Species* (New York: Collier & Son, 1909), 174–75.
2. Victor Negus, *The Mechanism of the Larynx* (London: William Heinemann, 1929), v11.
3. Negus, *The Mechanism of the Larynx*, 14.
4. Konstantinos Markatos, et al., "Antoine Ferrein (1693-1796)—His Life and Contribution to Anatomy and Physiology: The Description of the Vocal Cords and Their Function," *Surg Innov* 26, no. 3 (2019): 388–91, doi:10.1177/1553350619835346.
5. Gunnar Broberg, "Classification of Man," in *Linnaeus: The Man and His Work*, ed. Tore Frangsmyr (Berkeley: University of California Press, 1983), 167.

THREE: EMOTION

1. Paul D. MacLean, *The Triune Brain in Evolution* (New York: Plenum Press, 1990).
2. *Current Topics in Primate Vocal Communication*, ed., Elke Zimmermann,

John D. Newman, and Uwe Jürgens (Springer Verlag, 1995). Quoting a German language study: M. Monnier and H. Willis, "Die integrative Tätigkeit des Nervensystems beim meso-rhombo-spinalen Anencephalus (Mittelhirnwesen 1953), Monatsschr. Psychiat. Neurol. 126: 239–273.

3. W. R. Hess, *Hypothalamus and Thalamus* (Stuttgart: Georg Thieme Verlag, 1969).
4. Uwe Jürgens and Detlev Ploog, "Cerebral Representation of Vocalization in the Squirrel Monkey," *Experimental Brain Research* 10 (1970): 532–54.
5. Charles Darwin, *The Expression of the Emotions in Man and Animals* (New York: D. Appleton & Co., 1886).
6. Eugene S. Morton, "On the Occurrence and Significance of Motivation-Structural Rules in Some Bird and Mammal Sounds," *American Naturalist* 3 (1977): 855–69.
7. John Ohala, "Sound Symbolism," http://www.linguistics.berkeley.edu/~ohala/papers/SEOUL4-sound_symbolism.pdf.
8. Antonio Damasio, *Descartes' Error: Emotion, Reason and the Human Brain* (New York: Penguin, 2005).
9. Sigmund Freud, *Civilization and Its Discontents* (New York: W. W. Norton, 2010).
10. Cornelius J. Werner et al., "Altered Amygdala Functional Connectivity in Adult Tourette's Syndrome," *European Archives of Psychiatry and Clinical Neuroscience* 260 (2010), Suppl 2:S95, S99, doi:10.1007/s00406-010-0161-7.
11. Steven Pinker, *The Stuff of Thought: Language as a Window into Human Nature* (New York: Penguin, 2008). Pinker also wrote about the limbic system's role in swearing in the article "What the F***," *The New Republic,* October 8, 2007, https://newrepublic.com/article/63921/what-the-f.
12. Edward Hitchcock and Valerie Cairns, "Amygdalotomy," *Postgraduate Medical Journal* 49 (December 1973): 897.
13. Not to be confused with *spasmodic dysphonia*, a congenital speech disorder originating in the brain's motor centers, particularly those that activate opening and closing of the vocal cords for phonation. The vocal cords permanently slam together, producing a choked, wobbling, stop-start quality to the voice. The Kennedy clan has the gene for spasmodic dysphonia: you hear it in Robert Kennedy Jr.'s strangulated speech. Former NPR talk show host Diane Rehm is another famous sufferer—although she has a different version: her laryngeal spasms cause the vocal cords to stay *open*, preventing the proper chopping action of the airflow, creating a weak, breathy-sounding voice.

14. Jim Farber, "For Shirley Collins a Folk Revival of her Very Own," New York Times, November 7, 2016. Farber documents that Collins stayed silent as a singer, despite entreaties in the early 2000s from a new generation of fans, including Graham Coxon of the rock band Blur and Lee Ranaldo of Sonic Youth. However, in 2016, Collins, at eighty-two, recorded an album, *Lodestar*. Her voice, several octaves lower than the ethereal voice of her youth, has an austere, ragged tone that suits the bleak mood of the songs. Only in rare moments do her vocal cords seem to seize up, choking off a note prematurely.
15. Personal conversation with a leading voice surgeon who requested anonymity.
16. John R. Krebs and Richard Dawkins, "Animal Signals: Mindreading and Manipulation," *Behavioural Ecology, An Evolutionary Approach,* eds. Krebs and Dawkins (London: Blackwell Scientific Publications, 1984) 380–402.
17. Author interview with Klaus Scherer, May 9, 2018.
18. Paul Ekman, *Emotions Revealed* (New York: Owl/Henry Holt, 2007), 6–12.
19. Author interview with Scherer.
20. Klaus R. Scherer, "Vocal Communication of Emotion: A Review of Research Paradigms," *Speech Communication* 40, no. 1–2 (April 2003): 227–56.
21. Gladys E. Lynch, "A Phonophotographic Study of Trained and Untrained Voices Reading Factual and Dramatic Material," *Archives of Speech* 1, no. 1 (1934): 9–25.
22. From a story by Dustin Hoffman that he now partly recants: http://www.legacy.com/news/celebrity-deaths/article/the-quotable-laurence-olivier.
23. G. B. Duchenne de Boulogne, *The Mechanism of Human Facial Expression*, trans. A. Cuthbertson (Cambridge: Cambridge University Press, 1990), quoted in Ekman, *Emotions Revealed.*
24. Scherer, "Vocal Communication of Emotion: A Review of Research Paradigms," 227–56.
25. Klaus Scherer and Rainer Banse, "Acoustic Profiles in Vocal Emotion Expression," *Journal of Personality and Social Psychology* 70, no. 3 (1996): 614–36.
26. R. W. Picard, "Affective Computing," MIT Technical Report #321 (1995), https://affect.media.mit.edu/pdfs/95.picard.pdf.
27. https://www.youtube.com/watch?v=_86GQiEOjp4.
28. Picard, "Affective Computing," 2.
29. Author interview with Björn Schuller, August 2018.
30. Steven Pinker, *The Language Instinct* (New York: HarperPerennial 2007). This edition includes an essay that updates the original 1995 book and notes, on page 14 of the added material, how speech technology had advanced "tremendously."

31. Labeling emotions remains the most significant stumbling block, Schuller explained to me. But he, and others, are learning to get around it.
32. Picard, "Affective Computing," 8.
33. Picard, "Affective Computing," 15.

FOUR: LANGUAGE

1. M. Christiansen and S. Kirby, "Language Evolution: Consensus and Controversies," *Trends in Cognitive Science* 7 (2003): 300–307, quoted by the evolutionary biologist Tecumseh Fitch in his superb book *The Evolution of Language* (New York: Cambridge University Press, 2010), 15. Fitch, whom I came to know personally during our week together in the Amazon in 2007, helped spur my interest in language origins, and his book was indispensable to me.
2. Étienne Bonnot de Condillac, *Essay on the Origin of Human Knowledge* (1746).
3. Max Müller, *Lectures on the Science of Language* (New York: Charles Scribner, 1862).
4. Müller's theory—later parodied by rivals as the "Ding Dong theory"—sought to trace, in ancient Indo-European tongues, the roots of the biblical Adam's names for the animals.
5. Müller, *Lectures on the Science of Language*, 354.
6. A. R. Wallace, "The Development of the Human Races Under the Law of Natural Selection," *Journal of the Anthropological Society of London* 2 (1864), clviii–clxxxvii.
7. Darwin actually draws a parallel between the hand and the vocal organs: "One can hardly doubt, that a man-like animal who possessed a hand and arm sufficiently perfect to throw a stone with precision, or to form a flint into a rude tool, could, with sufficient practice, as far as mechanical skill alone is concerned, make almost anything which a civilized man can make. The structure of the hand in this respect may be compared with that of the vocal organs, which in the apes are used for uttering various signal-cries, or, as in one genus, musical cadences; but in man the closely similar vocal organs have become adapted through the inherited effects of use for the utterance of articulate language." *The Descent of Man and Selection in Relation to Sex* (New York: D. Appleton & Co., 1882), 50.
8. Darwin, *The Descent of Man and Selection in Relation to Sex*, 87.
9. Darwin, *The Descent of Man and Selection in Relation to Sex*, 87.
10. Darwin, *The Descent of Man and Selection in Relation to Sex*, 87.
11. Darwin, *The Descent of Man and Selection in Relation to Sex*, 88. Darwin actually

references Dr. Frederick Bateman's book *On Aphasia* (1870), which drew extensively on Broca's research with stroke patients.

12. Edward Sapir, "Herder's Ursprung der Sprache" (1907), 2, https://www.journals.uchicago.edu/doi/pdfplus/10.1086/386734.
13. Sapir, "Herder's Ursprung der Sprache," 142.
14. Edward Sapir, *Language* (New York: Harcourt Brace, 1921), 21–22.
15. Waldemar Kaempffert, "Science in Review," *New York Times*, June 18, 1950, 104.
16. B. F. Skinner, *Verbal Behavior* (New York: Appleton-Century-Crofts, 1957).
17. Skinner, *Verbal Behavior*, 462–63. "The selection of an instinctive response by its effect in promoting the survival of a species resembles, except for enormous differences in time scales, the selection of a response through reinforcement."
18. Noam Chomsky, "A Review of Skinner's *Verbal Behavior*," in *Language* 35, no. 1 (1959): 26–58. The entire eviscerating takedown is posted online: https://chomsky.info/1967____/.
19. Noam Chomsky, *Language and Mind* (New York: Harcourt, Brace & World, 1972), 97. Of the belief that language arose from Darwinian natural selection, Chomsky wrote that there is "no substance to this assertion."
20. Noam Chomsky, *Reflections on Language* (New York: Pantheon 1975), 59.
21. Chomsky floated this idea, somewhat tentatively, in his early book *Reflections on Language* (New York: Pantheon Books, 1975), 55–57. In response to strong pushback from notable scientists, Chomsky has only grown more outspoken, not to say belligerent, on this point, stating, in a 2013 lecture posted to Youtube, that language arrived "as an instrument of thought," and that "the modern dogmas about language and communication are just wrong," https://www.youtube.com/watch?v=iR_NmkkMmO8&feature=youtu.be, 16:48-17:04.
22. Steven Pinker has applied the critical term "guru" to Chomsky often, including in *The Language Instinct* (New York: Harper Perennial Modern Classics, 2007), 11, and in an interview with me, "The Interpreter," *The New Yorker,* April 16, 2007, 131.
23. Christine Kenneally, *The First Word* (New York: Penguin, 2007), 70.
24. Later work by Lieberman and colleagues showed that apes and monkeys can make other schwa-range vowels: those in "bit," "bet," "bat," "but," and "bought," which expand the repertoire of nonhuman primate sounds, but not enough for intelligible language, since they omit the all-important "point vowels," eee, ah,

ooo. Philip Lieberman, Edmund S. Crelin, Dennis H. Klatt, "Phonetic Ability and Related Anatomy of the Newborn and Adult Human, Neanderthal Man, and Chimpanzee," *American Anthropologist*, 74 (1972), 287-307.]

25. Lieberman's student, Tecumseh Fitch, would offer fascinating insight into why the proto-human larynx would have descended even before speech evolved: the bigger throat resonator increased the threat sound in a size-bluffing shout or growl—a highly useful adaptation in so puny, weak, slow-running and easily preyed-upon a creature. Fitch also showed that some animals, like the red deer, also evolved a permanently descended larynx, presumably for the same "size bluff" reasons. Tecumseh Fitch, "Comparative vocal production and the evolution of speech: Reinterpreting the descent of the larynx, in *The Transition to Language*, ed. A. Wray (Oxford: Oxford University Press, 2002), 21–45.
26. Charles Darwin, *The Origin of Species* (New York: Collier & Son, 1909), 176.
27. Edmund S. Crelin, *Anatomy of the Newborn: An Atlas* (Philadelphia: Lea & Febiger, 1969).
28. Detlev Ploog, "The Evolution of Vocal Communication," in *Nonverbal Vocal Communication,"* eds. Hanus Papousek, Uwe Jurgens, and Mechtild Papousek (Cambridge: Cambridge University Press, 1992), 17.
29. Edmund S. Crelin, *The Human Vocal Tract: Anatomy, Function, Development and Evolution* (New York: Vantage Press, 1987).
30. Philip Lieberman and Robert McCarthy, "Tracking the Evolution of Language and Speech," *Expedition* 49, no. 2 (2007): 17.
31 The tiny trace of Neanderthal DNA found in modern African populations is from the humans who returned to Africa, bringing some Neanderthal genes with them.
32. Lieberman sites a 1964 Harvard University photo facsimile of Charles Darwin, *The Origin of Species* (London: John Murray, 1859), 110
33. Philip Lieberman, *The Biology and Evolution of Language* (Cambridge: Harvard University Press, 1984), 329.
34. Lieberman, *The Biology and Evolution of Language*, 34, 35, 225, 331.
35. Philip Lieberman, B.G. Kanki, A. Protopappas, E. Reed and J.W. Youngs, "Cognitive Deficits at Altitude," *Nature* 372 (December 1994) 325.
36. Myra Gopnik, "Feature Blind Grammar and Dysphasia," *Nature* 344 (April 19, 1990): 715.
37. Pinker, *The Language Instinct*, 302–39.

38. Faraneh Vargha-Khadem et al., "Praxic and Nonverbal Cognitive Deficits in a Large Family with a Genetically Transmitted Speech and Language Disorder," *Proceedings of the National Academy of Science* 92 (January 1995): 930–33.
39. Faraneh Vargha-Khadem et al., "Neural Basis of an Inherited Speech and Language Disorder," *Proceedings of the National Academy of Science* 95 (October 1998): 12695–700.
40. Faraneh Vargha-Khadem et al., "Neural Basis of an Inherited Speech and Language Disorder," *Proceedings of the National Academy of Science* 95 (October 1998): 12695.
41. Wolfgang Enard et al., "Molecular Evolution of *FOXP2,* a Gene Involved in Speech and Language," *Nature* 418 (August 2002): 869–72.
42. Wolfgang Enard, et al., "A Humanized Version of FOXP2 Affects Cortico-basal Ganglia Circuits in Mice," *Cell,* 137, no. 5 (2009), 961–71.
43. S. Haesler et al., "Incomplete and Inaccurate Vocal Imitation After Knock-down of *FoxP2* in Songbird Basal Ganglia Nucleus Area X," *PLOS Biology* (December 4, 2007). When these researchers used targeted viruses to interrupt FOXP2's expression in the brain of baby zebra finches during the period when they learn their songs, the birds could not properly sing, despite concerted tutoring by adult finches.
44. Lieberman and McCarthy, "Tracking the Evolution of Language and Speech," 16.
45. The interview appeared in the free handout newspaper for the homeless, *Spare Change News,* 1999.
46. Marc D. Hauser, Noam Chomsky, and W. Tecumseh Fitch, "The Faculty of Language: What Is It, Who Has It, and How Did It Evolve?," *Science* 298 (November 22, 2002): 1569–79.
47. Ray Jackendoff and Steven Pinker, "The Nature of the Language Faculty and Its Implications for Evolution of Language (Reply to Fitch, Hauser, and Chomsky)," *Cognition* 97 (2005): 211–25.
48. Daniel L. Everett, "Cultural Constraints on Grammar and Cognition in Pirahã," *Current Anthropology* 46, no. 4 (August–October 2005): 621–46.
49. Author interview with Daniel Everett, January 2007.
50. Everett, "Cultural Constraints on Grammar and Cognition in Pirahã," 644.
51. Author interview with Brent Berlin, June 29, 2006.
52. John Colapinto, "The Interpreter," *The New Yorker,* April 16, 2007, 118–37. The rest of this chapter draws on the research I did for the article.
53. Unfortunately, Fitch has never published the results of these tests.

FIVE: SEX AND GENDER

1. Tom Gamill and Max Pross, "The Pledge Drive," *Seinfeld*, Season 6, Episode 3, original airdate, October 6, 1994.
2. Anthropologist David Puts in a YouTube lecture at the Leakey Foundation: "Being Human," https://www.youtube.com/watch?v=h8jsR8u2y9w, uploaded November 1, 2016.
3. Sedaris is interviewed in the movie *Do I Sound Gay?*
4. J. Oates and G. Dacakis, "Voice Change in Transsexuals," *Venereology* 10 (1997): 178–87. The authors measured typical frequency ranges: male voices in a range of 80–165 Hz; female voices 145–275 Hz with an overlap from 145 Hz to 165 Hz where the fundamental frequency cannot be assigned to one gender uniquely.
5. David A. Puts, Leslie M. Doll, and Alexander K. Hill, "Sexual Selection on Human Voices," *Evolutionary Perspectives on Human Sexual Psychology and Behavior*, eds. Viviana A. Weekes-Shackelford and Todd K. Shackelford (New York: Springer, 2014), 69–86.
6. Puts, Doll, and Hill, "Sexual Selection on Human Voices," 70.
7. J. S. Jenkins, "The Voice of the Castrato," *The Lancet* 351 (1998): 1877–80.
8. Jenkins, "The Voice of the Castrato," quoting C. De Brosses, "Lettres historiques et critiques sur l'Italie," Vol. 3 (Paris, 1799), 246.
9. H. Pleasants, "The Castrati," *Stereo Review*, July 1966, 38.
10. Darwin, *The Origin of the Species*, (London: John Murray, 1859), 88.
11. David Puts, S. J. C. Gaulin, and K. Verdolini, "Dominance and the Evolution of Sexual Dimorphism in Human Voice Pitch," *Evolution and Human Behavior* 27, no. 4 (2006): 283–96.
12. David Puts, "Mating Context and Menstrual Phase Affect Women's Preferences for Male Voice Pitch," *Evolution and Human Behavior* 26 (2005): 388–97.
13. Puts, "Mating Context and Menstrual Phase Affect Women's Preferences for Male Voice Pitch."
14. To say nothing of all the other negative social effects related to testosterone, including violence, divorce, and low investment in mates and offspring. A. Booth and J. M. Dabbs, "Testosterone and Men's Marriages," *Social Forces* 72 (1993): 463–77; and T. Burnham et al., "Men in Committed, Romantic Relationships Have Lower Testosterone," *Hormones and Behavior* 44 (2003): 119–22.
15. Puts, "Being Human."
16. Richard O. Prum, *The Evolution of Beauty* (New York: Anchor, 2018).
17. Geoffrey F. Miller, "Evolution of Human Music Through Sexual Selection," in

N. L. Wallin, B. Merker, and S. Brown, eds., *The Origins of Music* (Cambridge: MIT Press, 2000), 329–60.

18. Daniel J. Levitin, *This Is Your Brain on Music: The Science of a Human Obsession* (New York: Dutton, 2016), 252.
19. P. J. Fraccaro, B. C. Jones, J. Vukovic, F. G. Smith, C. D. Watkins, et al., "Experimental Evidence That Women Speak in a Higher Voice Pitch to Men They Find Attractive," *Journal of Evolutionary Psychology* 9 (2011): 57–67.
20. D. R. Feinberg, "Are Human Faces and Voices Ornaments Signaling Common Underlying Cues to Mate Value?," *Evolutionary Anthropology* 17 (2008): 112–18; and S. M. Hughes, F. Dispenza, and G. G. J. Gallup, "Ratings of Voice Attractiveness Predict Sexual Behavior and Body Configuration," *Evolution and Human Behavior* 25 (2004): 295–304; and Yi Xu et al., "Human Vocal Attractiveness as Signaled by Body Size Projection," *PLOS ONE* (April 24, 2013), https://journals.plos.org/plosone/article?id=10.1371/journal.pone.0062397.
21. Maria Südersten and Per-Åke Lindestad, "Glottal Closure and Perceived Breathiness During Phonation in Normally Speaking Subjects," *Journal of Speech and Hearing Research* 33 (1990): 601–11.
22. Reneé Van Bezooijen, "Sociocultural Aspects of Pitch Differences Between Japanese and Dutch Women," *Language and Speech* 38, no. 3 (1990): 253–65.
23. Nalina Ambady et al., "Surgeons' Tone of Voice: A Clue to Malpractice History," *Surgery* 132, no. 1 (2002): 5–9.
24. Cecilia Pemberton, Paul McCormack, and Alison Russell, "Have Women's Voices Lowered Across Time? A Cross Sectional Study of Australian Women's Voices," *Journal of Voice* 12, no. 2 (June 1998): 208–13.
25. Tina Tallon, "A Century of 'Shrill': How Bias in Technology Has Hurt Women's Voices," *The New Yorker,* September 3, 2019, https://www.newyorker.com/culture/cultural-comment/a-century-of-shrill-how-bias-in-technology-has-hurt-womens-voices.
26. C. E. Linke, "A Study of Pitch Characteristics of Female Voices and Their Relationship to Vocal Effectiveness," *Folia Phoniat* 25 (1973): 173–85.
27. Maria DiBattista, *Fast-Talking Dames* (New Haven: Yale University Press, 2001).
28. Lauren Bacall, *By Myself* (New York: Alfred A. Knopf, 1978).
29. Richard Brody, "The Shadows of Lauren Bacall," *The New Yorker*, August 13, 2014, https://www.newyorker.com/culture/richard-brody/shadows-lauren-bacall.

30. Germaine Greer, "Siren Song," *The Guardian,* December 30, 2006, https://www.theguardian.com/film/2006/dec/30/film.
31. Gloria Steinem, *Outrageous Acts and Everyday Rebellions* (New York: Signet, 1983), 211.
32. Mary Beard, *Women and Power: A Manifesto* (New York: Liveright, 2017), 3.
33. Tonja Jacobi and Dylan Schweers, "Justice, Interrupted: The Effect of Gender, Ideology and Seniority at Supreme Court Oral Arguments," *Virginia Law Review* 103 (March 2017): 1379–1496.
34. Rebecca Solnit, *Men Explain Things to Me* (Chicago: Haymarket Books, 2014).
35. Solnit, *Men Explain Things to Me*, 4.
36. Ikuko Patricia Yuasa, "Creaky Voice: A New Feminine Voice Quality for Young Urban-Oriented Upwardly Mobile American Women?," *American Speech* 85, no. 3 (2010): 315–37.
37. Mark Liberman, "Freedom Fries," *Language Log,* February 3, 2015, https://languagelog.ldc.upenn.edu/nll/?p=17489.
38. Yuasa, "Creaky Voice."
39. Rindy C. Anderson and Casey A. Klofstad, "Vocal Fry May Undermine the Success of Young Women in the Labor Market," *PLOS ONE* 9(5): e97506, https://doi.org/10.1371/journal.pone.0097506.
40. Tom Wolfe, *The Right Stuff* (New York: Bantam, 2001), 33–35.
41. https://www.quora.com/Is-Chuck-Yeager-voice-something-they-teach-in-flight-school-Why-do-all-airline-pilots-have-the-same-cool-calm-cadence-when-they-speak.
42. "US Airways Flight 1549 Full Cockpit Recording," YouTube, uploaded February 5, 2009, https://www.youtube.com/watch?v=mLFZTzR5u84.
43. Lynette Rice, "'Keeping Up with the Kardashians' Premiere Attracts Record Audience," *Entertainment Weekly*, August 23, 2010.
44. Consistent with this are the findings of Penelope Eckert, a leading expert in the sociological ramifications of how we use the voice, who told NPR's Ira Glass that the young women in her Stanford class thought the vocal fry in NPR announcers' voices sounded "authoritative"—whereas Eckert, who is in her sixties, thought the opposite. "I knew I was behind the curve," she told Glass. Liberman, "Freedom Fries."
45. Mark Liberman, "You Want Fries with That?," *Language Log,* posted February 3, 2015, http://languagelog.ldc.upenn.edu/nll/?p=17496.

46. DVD extras, *Seinfeld Season 6: Notes About Nothing*, "The Pledge Drive," Sony Pictures Home Entertainment, 2005.
47. S. E. James, J. L. Herman, S. Rankin, M. Keisling, L. Mottet, and M. Anafi, "The Report of the 2015 U.S. Transgender Survey" (Washington, DC: National Center for Transgender Equality, December 2016), https://transequality.org/sites/default/files/docs/usts/USTS-Full-Report-Dec17.pdf.
48. For an excellent summary on transwomen vocal surgery I recommend "Care of the Transgender Voice: Focus on Feminization," UCLA Gender Health, on YouTube: https://www.youtube.com/watch?v=sGxMA8JMBj0, uploaded December 18, 2018.
49. Stef Sanjati, "Voice Training 101 (for Trans Women)," YouTube, uploaded April 21, 2016, https://www.youtube.com/watch?v=q6eTvS2wIUc&t=640s.
50. Janet Pierrehumbert, "The Influence of Sexual Orientation on Vowel Production," *Journal of the Acoustical Society of America* 116, no. 4 (October 2004): 1905–8.
51. Sue Ellen Linville, "Acoustic Correlates of Perceived Versus Actual Sexual Orientation in Men's Speech," *Folia Phoniatrica et Logopaedica* 50, no. 1 (1998): 35–48.
52. Linville, "Acoustic Correlates of Perceived Versus Actual Sexual Orientation in Men's Speech": "gay judgments were significantly associated with higher peak /s/ frequency values and longer /s/ duration values." Also: Rudolf P. Gaudio, "Sounding Gay: Pitch Properties in the Speech of Gay and Straight Men," *American Speech* 69, no. 1 (1994): 30–57.
53. Janet B. Pierrehumbert, Tessa Bent, et al., "The Influence of Sexual Orientation on Vowel Production," *Journal of the Acoustical Society of America* 116, no. 4, part 1 (October 2004): 1905–8.
54. Pierrehumbert and Bent, "The Influence of Sexual Orientation on Vowel Production": "young people predisposed to becoming GLB adults (perhaps through a genetic disposition or difference in prenatal environment) selectively attend to certain aspects of opposite-sex adult models during early language acquisition" (1908).
55. Pierrehumbert and Bent, "The Influence of Sexual Orientation on Vowel Production," 1908.
56. *Do I Sound Gay?*, director, David Thorpe, release date, July 10, 2015.
57. Guy Branum, *My Life as a Goddess: A Memoir Through (Un)Popular Culture* (New York: Atria, 2019), 49.

58. John Laver, "Phonetic and Linguistic Markers in Speech," in *The Gift of Speech: Papers in the Analysis of Speech and Voice* (Edinburgh: Edinburgh University Press, 1991), 246.

SIX: THE VOICE IN SOCIETY

1. George Bernard Shaw, preface, *Pygmalion* (London: Penguin, 2003), 3.
2. Patricia E. G. Bestelmeyer, Pascal Belin, and D. Robert Ladd, "A Neural Marker for Social Bias Toward In-Group Accents," *Cerebral Cortex* 25, no. 10 (October 2015): 3953–61, https://doi.org/10.1093/cercor/bhu282.
3. Jairo N. Fuertes, William H. Gottdiener, et al., "A Meta-Analysis of the Effects of Speakers' Accents on Interpersonal Evaluations," *European Journal of Social Psychology* 42 (2012): 120–33.
4. Howard Giles and Caroline Sassoon, "The Effect of Speaker's Accent, Social Class Background and Message Style on British Listeners' Social Judgements," *Language & Communication* 3 (1983): 305–13.
5. Robert McCrum, William Cran, and Robin MacNeil, *The Story of English* (New York: Viking, 1986), 21.
6. Bill Bryson, *The Mother Tongue and How It Got That Way* (New York: William Morrow, 1990), 109.
7. Thomas Sheridan, *British Education: Or, The Source of the Disorders of Great Britain* (London: R. and J. Dodsley, 1757).
8. Thomas Sheridan, *A Course of Lectures on Elocution* (London: W. Strahan, 1753), 30.
9. Sheridan, *A Course of Lectures on Elocution*, 30.
10. McCrum, Cran, and MacNeil, *The Story of English*.
11. McCrum, Cran, and MacNeil, *The Story of English*.
12. Raymond Williams, *The Long Revolution* (London: Chatto & Windus, 1961), 247.
13. John Honey, *Tom Brown's Universe: The Development of the English Public School in the Nineteenth Century* (New York: Quadrangle, 1977), 233.
14. Honey, *Tom Brown's Universe*, 233.
15. McCrum, Cran, and MacNeil, *The Story of English*, 24.
16. Vivian Ducat, "Bernard Shaw and King's English," *Shaw* 9 (1989): 186.
17. John Reith, *Broadcast over Britain* (London: Hodder & Stoughton, 1924).
18. Shaw said that he became interested in phonetics "toward the end of the

eighteen seventies," which means he could have been as young as twenty. A relentless self-improver, it is hard to imagine that his "interest" in the subject did not include ridding himself of his extremely out-group Dublin accent. None of Shaw's innumerable biographers have addressed the question, which, to my mind, demonstrates an amazing incuriosity, at the very least.

19. Ducat, "Bernard Shaw and King's English," 187.
20. Ducat, "Bernard Shaw and King's English," 190.
21. Daniel Jones, *An English Pronouncing Dictionary* (London: J. M. Dent & Sons, sixth ed., 1944), x–x1.
22. Afferbeck Lauder, *Fraffley Suite* (London: Ure Smith/Wolfe, 1969), 13.
23. F. Scott Fitzgerald, *The Great Gatsby* (New York: Scribner, 1995), 127.
24. Fitzgerald, *The Great Gatsby*, 13–14.
25. Fitzgerald, *The Great Gatsby*, 11.
26. Author interview with William Labov, April 7, 2017.
27. William Labov, "The Social Motivation of a Sound Change," *Word* 19, no. 3 (1963): 273–309.
28. William Labov, *The Social Stratification of English in New York* (Cambridge: Cambridge University Press, 2006).
29. Edward McClelland, *How to Speak Midwestern* (Cleveland: Belt Publishing, 2016).
30. Edward Hall Gardner and Edwin Ray Skinner, *Good Taste in Speech: The Manual of Instruction of the Pronunciphone Course* (Chicago: Pronunciphone Company, 1928).
31. William Labov, *Dialect Diversity in America: The Politics of Language Change* (Charlottesville: University of Virginia Press, 2009).
32. McClelland, *How to Speak Midwestern*, 23.
33. The historical-political account that follows is from Labov's *Dialect Diversity in America.*
34. This comes from Labov's quotation, in *Dialect Diversity in America,* of *The History of McLean County, Illinois* (Chicago: Wm. Le Baron, Jr. & Co: 1879), 97.
35. Labov, *Dialect Diversity in America*, 38.
36. John Russell Rickford, *African American Vernacular English: Features and Use* (Hoboken, NJ: Wiley-Blackwell, 1999).
37. Author interview with John Baugh, April 19, 2017.
38. T. Purnell, W. Idsari, and John Baugh, "Perceptual and Phonetic Experiments

on American English Dialect Identification," *Journal of Language and Social Psychology* 18 (1999): 10–30.

39. William Labov, "The Logic of Nonstandard English," in *Report of the Twentieth Annual Round Table Meeting on Linguistics and Language Studies* (Washington, DC: Georgetown University Press, 1969), 1–44.
40. C. Bereiter and S. Engelmann, *Teaching Disadvantaged Children in the Preschool* (Englewood Cliffs, NJ: Prentice-Hall, 1966).
41. John Russell Rickford and Russell John Rickford, *Spoken Soul: The Story of Black English* (Hoboken, NJ: Wiley, 2000), 195.
42. John McWhorter, *Talking Back, Talking Black* (New York: Bellevue Literary Press, 2017), 15.
43. Rickford and Rickford, *Spoken Soul*, 147–52.
44. John Gramlich, "Black Imprisonment Rate in the U.S. Has Fallen by a Third Since 2006," *Fact Tank*, Pew Research Center (May 6, 2020), https://pewrsr.ch/2zc6PKi.
45. Rickford and Rickford, *Spoken Soul*, 223, citing Signithia Fordham and John Ogbu, "Black Students' School Success: Coping with the Burden of Acting White," *Urban Review* (1986): 181–82.
46. Labov, *Dialect Diversity in America.*
47. Author interview with John McWhorter, June 2017.
48. McWhorter, *Talking Back, Talking Black*, 71.
49. McWhorter, *Talking Back, Talking Black*, 73
50. Or longer. According to Marvin McAllister, author of *Whiting Up: Whiteface Minstrels and Stage Europeans in African American Performance* (Chapel Hill: University of North Carolina Press, 2014), nineteenth-century African American "white minstrels" "appropriated white-identified gestures, vocabulary, dialects, dress, or social entitlements . . . [to] satirize, parody, and interrogate privileged or authoritative representations of whiteness." Page 1.
51. William Labov, Sharon Ash, and Charles Boberg, *The Atlas of North American English* (New York: Mouton de Gruyter, 2005).
52. Labov, *Dialect Diversity in America.*
53. Labov, *Dialect Diversity in America.*

SEVEN: THE VOICE OF LEADERSHIP AND PERSUASION

1. Thomas Habinek, "Introduction," *Ancient Rhetoric from Aristotle to Philostratus*, trans. and ed. Thomas Habinek (New York: Penguin, 2017), xi.

2. Habinek, "Introduction," xviii.
3. Ben Yagoda, *The Sound on the Page: Style and Voice in Writing* (New York: Harper-Resource, 2004), 6–7.
4. G. Blakemore Evans, "Shakespeare's Text," *The Riverside Shakespeare,* 2nd ed (Boston: Houghton Mifflin, 1997), 55–69.
5. Vladimir Nabokov, *Lolita* (New York: G. P. Putnam's Sons, 1955), 11.
6. Carolyn Eastman, "Oratory and Platform Culture in Britain and North America, 1740–1900" (Oxford Handbooks Online, July 2016), 7.
7. Marlana Portolano, *The Passionate Empiricist* (Albany: State University of New York Press, 2009), 27.
8. Eastman, "Oratory and Platform Culture in Britain and North America, 1740–1900," 8.
9. Eastman, "Oratory and Platform Culture in Britain and North America, 1740–1900," 19.
10. Horace White, *The Lincoln and Douglas Debates: An Address Before the Chicago Historical Society* (Chicago: Chicago Historical Society, 1914), 20.
11. Wendi Maloney, "Hearing Abraham Lincoln's Voice," Library of Congress blog (January 3, 2018), https://blogs.loc.gov/loc/2018/01/hearing-abraham-lincolns-voice/.
12. https://www.youtube.com/watch?v=5g9v8y5FvSo, December 3, 2012.
13. White, *The Lincoln and Douglas Debates*, 20.
14. White, *The Lincoln and Douglas Debates*, 21.
15. Stephen A. Douglas, "Homecoming Speech at Chicago, July 9, 1858," https://teachingamericanhistory.org/library/document/homecoming-speech-at-chicago/.
16. "2018 Winter Lecture Series—The Lincoln-Douglas Debates," YouTube, uploaded June 14, 2018, https://www.youtube.com/watch?v=0NgmkFy5EJM.
17. History.com, Editor, "President Lincoln Delivers Gettysburg Address," https://www.history.com/this-day-in-history/lincoln-delivers-gettysburg-address.
18. William Strunk, Jr., and E. B. White, *The Elements of Style* (New York: Longman, 2000), 77.
19. Gesine Manuwald, *Cicero* (New York: Bloomsbury, 2014), 142.
20. Richard Hidary, "Rabbis and Classical Rhetoric," in *Rabbis and Classical Rhetoric: Sophistic Education and Oratory in the Talmud and Midrash* (Cambridge: Cambridge University Press, 2017), i–ii.
21. Philip Halldén, "What Is Arab Islamic Rhetoric? Rethinking the History of

Muslim Oratory Art and Homiletics," *International Journal of Middle East Studies* 37, no. 1 (February 2005): 23.

22. Halldén, "What Is Arab Islamic Rhetoric?," 22.
23. Jonathan Edwards, "Sinners in the Hands of an Angry God" (Boston: S. Kneeland and T. Green, 1741).
24. Jim Ehrhard, "A Critical Analysis of the Tradition of Jonathan Edwards as a Manuscript Preacher," *Westminster Theological Journal* (Spring 1998).
25. "HELL FIRE: The Most Powerful Sermon Ever!!!," YouTube, uploaded January 22, 2011, https://www.youtube.com/watch?v=fCnQQLUJHb8&t=2s.
26. Laurie F. Maffly-Kipp, "The Church in the Southern Black Community (May 2011), https://docsouth.unc.edu/church/intro.html, uploaded May 21, 2007.
27. Geneva Smitherman, *Talkin and Testifyin* (Detroit: Wayne State University Press, 1986).
28. Smitherman, *Talkin and Testifyin*, 134–35.
29. Jill Lepore, *These Truths: A History of the United States* (New York: W. W. Norton & Company, 2018).
30. Jill Lepore, *These Truths: A History of the United States*, 478.
31. History Matters website, http://historymatters.gmu.edu/d/8126.
32. I relied for the account that follows on Thomas Putnam, "The Real Meaning of *Ich Bin Ein Berliner*," in *Atlantic Monthly*, https://www.theatlantic.com/magazine/archive/2013/08/the-real-meaning-of-ich-bin-ein-berliner/309500/.
33. Author correspondence with V. S. Ramachandran, April 30, 2012.
34. Author interview with John Baugh, April 19, 2017. Baugh said of Obama: "When he is in a predominantly black situation he does a few things, and one that I noticed was that any multisyllabic word ending in the long 'e' sound is where he nails black phonology."
35. William Labov, *Dialect Diversity in America: The Politics of Language Change* (Charlottesville: University of Virginia Press, 2009).
36. H. Samy Alim and Geneva Smitherman, *Articulate While Black* (New York: Oxford University Press, 2012), 1–11.
37. David Remnick, *The Bridge* (New York: Alfred A. Knopf, 2010), 18.
38. Remnick, *The Bridge*, 361.
39. Adolf Hitler, *Mein Kampf* (Boston: Houghton Mifflin, 1943), 219.
40. Hitler, *Mein Kampf*, 55.
41. Hitler, *Mein Kampf*, 17.

42. Adolf Hitler, *My New Order*, edited with commentary by Raoul de Roussy de Sales (New York: Reynal & Hitchcock, 1941), xiv.
43. Joseph Goebbels, "Der Führer als Redner," *Adolf Hitler. Bilder aus dem Leben des Führers* (Hamburg: Cigaretten/Bilderdienst Hamburg/Bahrenfeld, 1936), 27–34, https://research.calvin.edu/german-propaganda-archive/ahspeak.htm.
44. Goebbels, "Der Führer als Redner."
45. Unsigned item, "Inside Hitler's Mind," on the University of Cambridge website (https://www.cam.ac.uk/research/news/inside-hitlers-mind), which also links to a copy of the original intelligence report, entitled "ANALYSIS OF HITLER'S SPEECH ON THE 26TH OF APRIL, 1942."
46. Ron Rosenbaum, *Explaining Hitler: The Search for the Origins of his Evil* (Boston: Da Capo, 2014), 303.
47. Hitler, *My New Order*, 3.
48. Annalisa Merelli, "The State of Global Right-Wing Populism in 2019," *Quartz* (December 30, 2019), https://qz.com/1774201/the-global-state-of-right-wing-populism-in-2019/.
49. These dismal facts come from a TED Talk lecture by the political scientist Abrak Saati: https://www.youtube.com/watch?v=fszENSaH0GQ&feature=youtu.be, uploaded June 7, 2018.
50. S. W. Gregory and T. J. Galeasher, "Spectral Analysis of Candidates' Nonverbal Vocal Communication: Predicting US Presidential Election Outcomes," *Social Psychology Quarterly* 65 (2002): 298–308.
51. Author interview with Branka Zei Pollerman, September 2, 2018. Zei Pollerman has done objective analysis of Trump's pitch and has compared it to George Clooney, who, she told me, "has a pitch of 65 Hz with a heavy vocal fry." Trump's pitch will sometimes jump as high as 147.4 Hz.
52. Jacquiline Sachs, Philip Lieberman, and D. Erickson, "Anatomical and Cultural Determinants of Male and Female Speech," *Language Attitudes: Current Trends and Prospects* (1972).
53. Ryan Dilbert, "Donald Trump: A History of the Presidential Candidate's Involvement with WWE," https://bleacherreport.com/articles/2669447-donald-trump-a-history-of-the-presidential-candidates-involvement-with-wwe.
54. Ryan Grim and Danny Shea, "A Note About Our Coverage of Donald Trump's 'Campaign,'" *Huffington Post* (July 17, 2015), https://www.huffpost.com/entry/a-note-about-our-coverage-of-donald-trumps-campaign_n_55a8fc9ce4b0896514d0fd66.

55. Arianna Huffington, "A Note on Trump: We Are No Longer Entertained," *Huffington Post* (December 7, 2015), https://www.huffpost.com/entry/a-note-on-trump_b_8744476.
56. Morris Kaplan, "Major Landlord Accused of Antiblack Bias," *New York Times*, October 16, 1973.
57. Jan Ransom, "Trump Will Not Apologize for Calling for Death Penalty over Central Park Five," *New York Times*, June 18, 2019.
58. Ransom, "Trump Will Not Apologize for Calling for Death Penalty over Central Park Five."
59. Marie Brenner, "After the Gold Rush," *Vanity Fair* (September 1990), 294.
60. Bob Woodward, *Fear: Trump in the White House* (New York: Simon & Schuster, 2019), 16.
61. Michael E. Miller, "Donald Trump on a Protester: 'I'd Like to Punch Him in the Face,'" *Washington Post*, February 23, 2016.
62. https://www.nbcnews.com/video/trump-tells-crowd-to-knock-the-crap-out-of-tomato-throwers-613684291706, February 1, 2016.
63. Nick Corasaniti and Maggie Haberman, "Donald Trump Suggests 'Second Amendment People' Could Act Against Hillary Clinton," *New York Times*, August 9, 2016.
64. Peter Osborne and Tom Roberts, *How Trump Thinks* (London: Head of Zeus, 2017), 207. Tweet of July 28, 2016.
65. Osborne and Roberts, *How Trump Thinks*, 198. Tweet of August 19, 2016.
66. Osborne and Roberts, *How Trump Thinks*, 198. Tweet of June 14, 2016.
67. *The Last Word with Lawrence O'Donnell*, MSNBC, November 8, 2017.
68. J. D. Vance, *Hillbilly Elegy: A Memoir of a Family and Culture in Crisis* (New York: Harper, 2016), 3.
69. Vance, *Hillbilly Elegy*, 191.
70. Michael D'Antonio, *The Truth About Trump* (New York: St. Martin's, 2016), 39.

EIGHT: SWAN SONG

1. Steven Pinker, *How the Mind Works* (New York: W. W. Norton, 1997), 534.
2. Pinker, *How the Mind Works*, 528.
3. Milton Metfessel and Carl E. Seashore, *Phonophotography in Folk Music: American Negro Songs in New Notation* (Chapel Hill: University of North Carolina Press, 1928).
4. Author interview with Robert Warsh, May 18, 2020.

5. Metfessel and Seashore, *Phonophotography in Folk Music*, 14.
6. The correlation between emotion and vibrato in singing was demonstrated in a recent study in *Journal of Voice* 29, no. 2 (March 2015): 170–81, "The Effects of Emotional Expression on Vibrato." Authors Christopher Dromey et al. recruited ten graduate school singers to sing songs with sustained vowels at several pitches and volumes. Songs personally selected by the singers for the intensity of the emotions resulted in stronger vibrato than the emotionally neutral songs picked by the researchers.
7. Johan Sundberg, "Acoustic and Psychoacoustic Aspects of Vocal Vibrato," *STL-QPSR* 35, no. 2–3 (1994): 45–68.
8. Carl E. Seashore, *Psychology of the Vibrato in Voice and Instrument* (Iowa City: University Press of Iowa, 1936).
9. Author interview with Ingo Titze, executive director of the National Center for Voice and Speech at the University of Utah in Salt Lake City, April 10, 2017.
10. Carl E. Seashore, *Psychology of the Vibrato in Voice and Instrument*, (1936), 111.
11. Carl E. Seashore, *Psychology of the Vibrato in Voice and Instrument*, (1936), 47.
12. Dr. Yusuf Dalhat, "The Concept of al-Ruh (Soul) In Islam," *International Journal of Education and Research* 3 no. 8 (August 2015): 431–40.
13. Hyun-Ah Kim, *The Renaissance Ethics of Music: Singing, Contemplation and Musica Humana* (Abingdon, UK: Routledge, 2015), 63.
14. Katherine Le Mée, *Chant: The Origin, Form, Practice and Healing Power of Gregorian Chant* (New York: Bell Tower, 1994), 98.
15. Lennon's twanging nasality is actually a feature of Liverpudlian speech that arises from holding the velum open slightly and sending the soundwave through the nose's muffling mucous membranes. Not all natives do this: George Harrison did, Paul and Ringo do not. Nevertheless, the sound is so unique to Liverpool that linguist Gerald Knowles has speculated that it resulted from the northern city's frigid winters and bad 19 century public health which gave rise to constant colds and partially congested noses, resulting in a twanging or "adenoidal" voice that was passed, by Motherese, even to healthy children. Gerald Knowles, "Scouse: The Urban Dialect of Liverpool" (PhD Thesis: University of Leeds, 1973), 116-117.
16. Nichola Gale, Stephanie Enright, et al., "A Pilot Investigation of Quality of Life and Lung Function Following Choral Singing in Cancer Survivors and Their Careers," *ecancer* 6, no. 261 (2012), doi:10.3332/ecancer.2012.261.

17. R. J. Beck, T. C. Cesario, et al., "Choral Singing, Performance Perception, and Immune System Changes in Salivary Immunoglobulin A and Cortisol," *Music Perception: An Interdisciplinary Journal* 18, no. 1 (Fall 2000): 87–106.
18. Rollin McCraty, Mike Atkinson, Glen Rein, and Alan D. Watkins, "Music Enhances the Effect of Positive Emotional States on Salivary IgA," *Stress Medicine* 12, no. 3 (1996): 167–75.
19. Jordyn Phelps, "The Story Behind President Obama Singing 'Amazing Grace' at Charleston Funeral," ABC News online, July 7, 2015, https://abcnews.go.com/Politics/story-president-obama-singing-amazing-grace-charleston-funeral/story?id=32264346. The story includes video of Jarrett's account.
20. "Spotlight," *New York Times,* July 6, 2017.
21. Renée Fleming, *The Inner Voice: The Making of a Singer* (New York: Viking, 2004), 20.
22. Fleming, *The Inner Voice*, 4.
23. Fleming, *The Inner Voice*, 17.
24. Author interview with Laurie Antonioli, November 22, 2016.
25. Charles Darwin, *The Descent of Man and Selection in Relation to Sex* (New York: D. Appleton & Co., 1882), 572.
26. Darwin, *The Descent of Man and Selection in Relation to Sex*, 572.

CODA

1. John Colapinto, "Giving Voice," *The New Yorker,* March 4, 2013, 48–57.
2. Emily Heil, "Chuck Leavell Throws a White House Correspondents' Dinner Pre-Party Where Journalists Rock," *Washington Post*, April 13, 2015.

INDEX